AF750547

Advances in Industrial Control

For other titles published in this series, go to
www.springer.com/series/1412

Other titles published in this series:

Digital Controller Implementation and Fragility
Robert S.H. Istepanian and James F. Whidborne (Eds.)

Optimisation of Industrial Processes at Supervisory Level
Doris Sáez, Aldo Cipriano and Andrzej W. Ordys

Robust Control of Diesel Ship Propulsion
Nikolaos Xiros

Hydraulic Servo-systems
Mohieddine Mali and Andreas Kroll

Model-based Fault Diagnosis in Dynamic Systems Using Identification Techniques
Silvio Simani, Cesare Fantuzzi and Ron J. Patton

Strategies for Feedback Linearisation
Freddy Garces, Victor M. Becerra, Chandrasekhar Kambhampati and Kevin Warwick

Robust Autonomous Guidance
Alberto Isidori, Lorenzo Marconi and Andrea Serrani

Dynamic Modelling of Gas Turbines
Gennady G. Kulikov and Haydn A. Thompson (Eds.)

Control of Fuel Cell Power Systems
Jay T. Pukrushpan, Anna G. Stefanopoulou and Huei Peng

Fuzzy Logic, Identification and Predictive Control
Jairo Espinosa, Joos Vandewalle and Vincent Wertz

Optimal Real-time Control of Sewer Networks
Magdalene Marinaki and Markos Papageorgiou

Process Modelling for Control
Benoît Codrons

Computational Intelligence in Time Series Forecasting
Ajoy K. Palit and Dobrivoje Popovic

Modelling and Control of Mini-Flying Machines
Pedro Castillo, Rogelio Lozano and Alejandro Dzul

Ship Motion Control
Tristan Perez

Hard Disk Drive Servo Systems (2nd Ed.)
Ben M. Chen, Tong H. Lee, Kemao Peng and Venkatakrishnan Venkataramanan

Measurement, Control, and Communication Using IEEE 1588
John C. Eidson

Piezoelectric Transducers for Vibration Control and Damping
S.O. Reza Moheimani and Andrew J. Fleming

Manufacturing Systems Control Design
Stjepan Bogdan, Frank L. Lewis, Zdenko Kovačić and José Mireles Jr.

Windup in Control
Peter Hippe

Nonlinear H_2/H_∞*, Constrained Feedback Control*
Murad Abu-Khalaf, Jie Huang and Frank L. Lewis

Practical Grey-box Process Identification
Torsten Bohlin

Control of Traffic Systems in Buildings
Sandor Markon, Hajime Kita, Hiroshi Kise and Thomas Bartz-Beielstein

Wind Turbine Control Systems
Fernando D. Bianchi, Hernán De Battista and Ricardo J. Mantz

Advanced Fuzzy Logic Technologies in Industrial Applications
Ying Bai, Hanqi Zhuang and Dali Wang (Eds.)

Practical PID Control
Antonio Visioli

(continued after Index)

Antonio Visioli · Qing-Chang Zhong

Control of Integral Processes with Dead Time

Springer

Prof. Antonio Visioli
Facoltà di Ingegneria
Dipartimento di Ingegneria
dell'Informazione
Università di Brescia
Via Branze 38
Brescia 25123
Italy
antonio.visioli@ing.unibs.it

Prof. Qing-Chang Zhong
Department of Aeronautical and Automotive
Engineering
Loughborough University
Loughborough, Leicestershire LE11 3TU
UK
Q.Zhong@lboro.ac.uk

ISSN 1430-9491
ISBN 978-0-85729-069-4 e-ISBN 978-0-85729-070-0
DOI 10.1007/978-0-85729-070-0
Springer London Dordrecht Heidelberg New York

British Library Cataloguing in Publication Data
A catalogue record for this book is available from the British Library

© Springer-Verlag London Limited 2011

Matlab® and Simulink® are registered trademarks of The MathWorks, Inc., 3 Apple Hill Drive, Natick, MA 01760-2098, USA. http://www.mathworks.com

Apart from any fair dealing for the purposes of research or private study, or criticism or review, as permitted under the Copyright, Designs and Patents Act 1988, this publication may only be reproduced, stored or transmitted, in any form or by any means, with the prior permission in writing of the publishers, or in the case of reprographic reproduction in accordance with the terms of licenses issued by the Copyright Licensing Agency. Enquiries concerning reproduction outside those terms should be sent to the publishers.
The use of registered names, trademarks, etc., in this publication does not imply, even in the absence of a specific statement, that such names are exempt from the relevant laws and regulations and therefore free for general use.
The publisher makes no representation, express or implied, with regard to the accuracy of the information contained in this book and cannot accept any legal responsibility or liability for any errors or omissions that may be made.

Cover design: eStudio Calamar S.L.

Printed on acid-free paper

Springer is part of Springer Science+Business Media (www.springer.com)

Advances in Industrial Control

Series Editors

Professor Michael J. Grimble, Professor of Industrial Systems and Director
Professor Michael A. Johnson, Professor (Emeritus) of Control Systems and Deputy Director

Industrial Control Centre
Department of Electronic and Electrical Engineering
University of Strathclyde
Graham Hills Building
50 George Street
Glasgow Gl 1QE
UK

Series Advisory Board

Professor E.F. Camacho
Escuela Superior de Ingenieros
Universidad de Sevilla
Camino de los Descubrimientos s/n
41092 Sevilla
Spain

Professor S. Engell
Lehrstuhl für Anlagensteuerungstechnik
Fachbereich Chemietechnik
Universität Dortmund
44221 Dortmund
Germany

Professor G. Goodwin
Department of Electrical and Computer Engineering
The University of Newcastle
Callaghan NSW 2308
Australia

Professor T.J. Harris
Department of Chemical Engineering
Queen's University
Kingston, Ontario
K7L 3N6
Canada

Professor T.H. Lee
Department of Electrical and Computer Engineering
National University of Singapore
4 Engineering Drive 3
Singapore 117576
Singapore

Professor (Emeritus) O.P. Malik
Department of Electrical and Computer Engineering
University of Calgary
2500, University Drive, NW
Calgary, Alberta
T2N 1N4
Canada

Professor K.-F. Man
Electronic Engineering Department
City University of Hong Kong
Tat Chee Avenue
Kowloon
Hong Kong

Professor G. Olsson
Department of Industrial Electrical Engineering and Automation
Lund Institute of Technology
Box 118
221 00 Lund
Sweden

Professor A. Ray
Department of Mechanical Engineering
Pennsylvania State University
0329 Reber Building
University Park
PA 16802
USA

Professor D.E. Seborg
Chemical Engineering
University of California Santa Barbara
3335 Engineering II
Santa Barbara
CA 93106
USA

Doctor K.K. Tan
Department of Electrical and Computer Engineering
National University of Singapore
4 Engineering Drive 3
Singapore 117576
Singapore

Professor I. Yamamoto
Department of Mechanical Systems and Environmental Engineering
Faculty of Environmental Engineering
The University of Kitakyushu
1-1, Hibikino,Wakamatsu-ku, Kitakyushu, Fukuoka, 808-0135
Japan

To our Parents

Series Editors' Foreword

The series *Advances in Industrial Control* aims to report and encourage technology transfer in control engineering. The rapid development of control technology has an impact on all areas of the control discipline. New theory, new controllers, actuators, sensors, new industrial processes, computer methods, new applications, new philosophies, ..., new challenges. Much of this development work resides in industrial reports, feasibility study papers and the reports of advanced collaborative projects. The series offers an opportunity for researchers to present an extended exposition of such new work in all aspects of industrial control for wider and rapid dissemination.

The abundance of low-level single-input, single-output loops in many industrial sectors, particularly the process industries (refining, chemicals, utilities, etc.), that can be controlled by simple controllers has guaranteed the enduring popularity of the proportional-integral-derivative (PID) control algorithm, and its software and hardware incarnations. Over the years, the academic control community has produced new developments for this type of controller, and it is possible to claim that an academic and industrial PID controller and implementation paradigm now exists. If milestones for this paradigm have to be identified, then three of them are (arguably):

- Ziegler and Nichols tuning rules (1942)
- Åström and Hägglund's relay-based tuning experiment (1985)
- Harris's performance assessment concepts (1989)

Each of the milestone achievements has been very influential, and each in turn has initiated at least a decade or more of academic investigation and industrial follow-up. Since 1992, the *Advances in Industrial Control* monograph series has been fortunate to publish interesting volumes in the PID control paradigm that continue the academic and industrial investigation into the theory and practice of PID control.

Integral processes with dead time form a group of process models that are widely found in industry, and it is not surprising that the PID control methodologies for model identification, controller design, stability determination, and implementation can be specialised for this process form. Antonio Visioli and Qing-Chang Zhong have published widely on PID control, and in this *Advances in Industrial Control* volume they focus successfully on the PID control specialisations that can be developed for integral processes with dead time.

The monograph first reports on what can be achieved in the framework of single-loop PID control. Thus, Part I includes model identification, a review of PID tuning rules, and interestingly, performance assessment and retuning for PID control with this special form of process model. Part II of the monograph considers some more advanced topics (such as feedforward control) in the framework of the two-degrees-of-freedom control structure. The use of the Smith predictor principle and its developmental variants occupies two chapters in this part of the monograph. On implementation, the text fields Chapter 5 on the "Plug & Play" idea, and Chapter 12 on some practical issues. The monograph will be of interest to a broad control readership, both industrial and academic, and forms a valuable new addition to the *Advances in Industrial Control* series.

Readers wishing to learn more about PID control may be interested to know that the series has published the following monographs in this field:

- *Autotuning of PID Controllers: Relay Feedback Approach* by Cheng-Ching Yu (ISBN 978-3-540-76250-8, 1999)
- *Advances in PID Control* by Tan Kok K., Wang Quing-Guo and Hang Chang C. with Tore J. Hägglund (ISBN 978-1-85233-138-2, 1999)
- *Structure and Synthesis of PID Controllers* by Aniruddha Datta, Ming-Tzu Ho and Shankar P. Bhattacharyya (ISBN 978-1-85233-614-1, 1999)
- *Practical PID Control* by Antonio Visioli (ISBN 978-1-84628-585-1, 2006)

In the related field of process performance assessment, the series has published the following monographs:

- *Performance Assessment of Control Loops: Theory and Applications* by Biao Huang and Sirish L. Shah (ISBN 978-1-85233-639-4, 1999)
- *Process Control Performance Assessment: From Theory to Implementation* by Andrzej Ordys, Damien Uduehi and Michael A. Johnson (Eds.) (ISBN 978-1-84628-623-0, 2007)

And for dead-time processes, the related series *Advanced Textbooks in Control and Signal Processing* has published:

- *Control of Dead-time Processes* by Julio E. Normey-Rico and Eduardo F. Camacho (ISBN 978-1-84628-828-9, 2007)

All of these texts are repeatedly referenced in books and journal and conference papers, and the Editors believe they form a significant collection of books for PID control systems research and practice. Consequently, it is a pleasure to add *Control of Integral Processes with Dead Time* by Antonio Visioli and Qing-Chang Zhong to this growing PID control library within the *Advances in Industrial Control* monograph series.

Industrial Control Centre
Glasgow
Scotland, UK

M.J. Grimble
M.A. Johnson

Preface

Integral processes with dead time are frequently encountered in engineering. Typical examples include tanks, where the level is controlled by manipulating the difference between the input and output flow rates, batch distillation columns, data communication networks, and supply chain management processes. Because they are not asymptotically stable (namely, they are not self-regulating), their control requires special attention, and for this reason, many control techniques have been proposed for this purpose. In particular, different approaches have been exploited to design PID controllers, which are by far the most widely adopted controllers in industry. However, it is recognised that in the presence of a large time delay in the process and of tight control requirements, a single degree-of-freedom PID controller may not suffice to obtain the desired performance. Indeed, in the last fifteen years, starting with the paper by Åström, Hang, and Lim in 1994, the research in the control of integral processes with dead time has become very active, mainly motivated by the fact that the well-known Smith predictor fails to provide a null steady-state error in the presence of a constant load disturbance. In this context, many two-degree-of-freedom schemes have been proposed, and their analysis and design have been discussed in a wide literature.

This book presents some of these techniques by fully characterising them from both academic and industrial points of view and highlighting the peculiarities of each of them. The control schemes and the procedures for the selection of the parameters are outlined clearly, and illustrative examples are presented in order to evaluate the rationale of each method and the performance achievable.

The book is divided into two parts: PID Control Schemes (Chapters 2 to 5) and Two-degree-of-freedom Control Schemes (Chapters 6 to 12). In the first part, the tuning of a PID controller and the determination of its stabilising region are addressed, in addition to a technique for the performance assessment (and retuning) of a PID controller and for the Plug&Control strategy. In the second part, different methodologies for the design of two-degree-of-freedom control schemes are presented, in particular those based on the Smith predictor concept. The achievable performance is quantitatively analysed, and some practical issues are dealt with.

The book can serve as a reference for postgraduate students and academic researchers. It will also help industrial practitioners solve their control problems by selecting the most suitable technique and by achieving the best cost/benefit ratio. Readers are assumed to know the fundamentals of linear control systems, which are typically acquired in a basic course in automatic control at the university level.

The authors would like to acknowledge all the people who have contributed to this book in one way or another, in particular, A. Piazzi, M. Veronesi, F. Padula, M. Beschi, L. Mirkin, H.X. Li, B. Wang, D. Rees, just to name a few. Special thanks go to the series Editors M. Grimble and M. Johnson, the Editor O. Jackson and the Editorial Assistant C. Cross for their help during the preparation of the manuscript. Partial support of the research of A. Visioli has been provided by the Italian Ministry for Universities and Research. A. Visioli thanks his beloved wife Silvia and his children Alessandra, Laura, and Andrea for their love and support during the time spent in writing the book. Q.-C. Zhong thanks his wife Shuhong for her endurance, love, support, and sacrifice for his research over the years and their daughters Lilly and Lisa for brightening his life. He would also like to thank the Royal Academy of Engineering and the Leverhulme Trust for the award of a Senior Research Fellowship (2009–2010).

University of Brescia, Italy — Antonio Visioli
Loughborough University, UK — Qing-Chang Zhong

Contents

List of Figures

List of Tables

Chapter 1
Introduction

Many industrial processes have the dynamics of an integrator plus dead time (IPDT) and are called integral processes with dead time. The most well-known examples are tanks, where the level is controlled by using the difference between the input and output flow rates as a manipulated variable, batch distillation columns, data communication networks, supply chain management processes, etc. These processes are not asymptotically stable (namely, they are not self-regulating), and therefore their control requires special attention. For this reason, many techniques have been proposed in the literature. In this chapter, some examples of integral processes with dead time are described, and then an overview of the book is given.

1.1 Examples of Integral Processes with Dead Time

1.1.1 Tanks with an Outlet

The most typical example of IPDT process is a storage (or surge) tank with an outlet pump [84], which is often extremely important for the successful operation of chemical plants, because it is through the proper control of flows and levels that the desired production rates and inventories are achieved [67]. Such a tank is shown in Figure 1.1, where h is the level of the liquid, Q_i is the input flow rate, and Q_o is the output flow rate set by the pump. Hence, it is independent of the liquid level (note that if the output flow is not actuated, then $Q_o(t) = \alpha\sqrt{2gh(t)}$, where g is the gravitational acceleration constant, and α is the cross-sectional area of the output orifice). A model of the process can be derived easily by considering the balance equation based on an instantaneous rate-of-change, namely, by considering that the rate-of-change of the total mass of liquid inside the tank is equal to the mass flow rate of the liquid into the tank minus the mass flow rate of the liquid out of the tank. Thus, by assuming that a perfect mixing in the tank is achieved, *i.e.*, that the density (mass/volume) of liquid in the tank does not depend on the position, then

$$\frac{dV(t)\rho}{dt} = Q_i(t)\rho_i - Q_o(t)\rho, \tag{1.1}$$

A. Visioli, Q.-C. Zhong, *Control of Integral Processes with Dead Time*,
Advances in Industrial Control,
DOI 10.1007/978-0-85729-070-0_1, © Springer-Verlag London Limited 2011

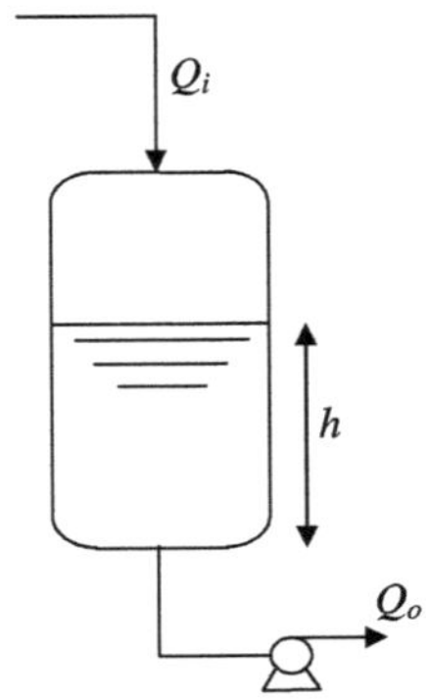

Fig. 1.1 A liquid-level control system

where V is the volume of the liquid in the tank, ρ is the density of the liquid in the tank, and ρ_i is the density of the inlet stream. Assuming that $\rho_i = \rho$ and that the cross-sectional area A of the tank is uniform, then

$$A\frac{dh(t)}{dt} = Q_o(t) - Q_i(t). \tag{1.2}$$

If the difference $Q_o(t) - Q_i(t)$ between the output and input flow rates is considered as the process input variable and the level h is the output variable, the process exhibits clearly an integral dynamic, that is, the solution of the differential equation (1.2) is given by

$$h(t) = h(0) + \frac{1}{A}\int_0^t \left[Q_o(v) - Q_i(v)\right] dv. \tag{1.3}$$

If, for example, the initial level is 2 metres, the cross-sectional area is 4 square metres, the inlet flow rate is 5 litres/second, and the outlet flow rate is 4 litres/seconds, then

$$h(t) = 2 + \frac{1}{4}t. \tag{1.4}$$

This means that a sudden change of the inlet flow rate while the outlet flow rate remains constant (or, similarly, a sudden change in the outlet flow rate while the inlet flow rate remains constant) results in a continuous increase or decrease in the liquid level. It is worth stressing that, in this context, Q_o can be selected as the manipulated input and Q_i is a disturbance input or *vice versa*. In any case, the open-loop process step response does not converge to a steady-state value, namely, the process is not self-regulating. In general, the transfer function of this kind of processes, which are also known as *pure capacity processes*, can be written as

$$P(s) = \frac{K}{s}. \tag{1.5}$$

In general, the presence of a (possibly apparent) dead time L can occur because of the length of the pipes between the valves and the tank, and/or because of the

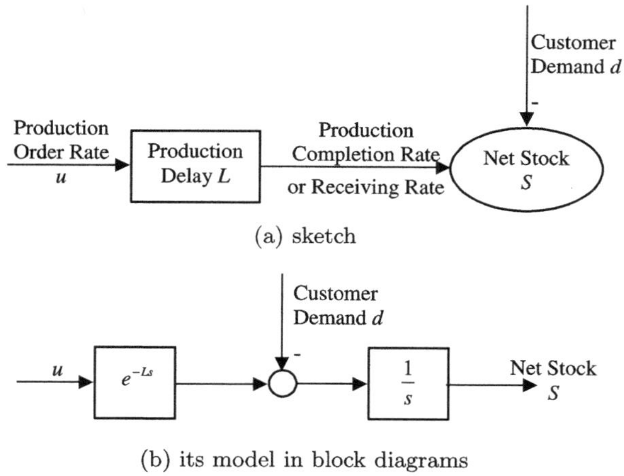

Fig. 1.2 A production process in supply chains

dynamics of the sensor and of the actuator. Thus, the transfer function of an IPDT process becomes

$$P(s) = \frac{K}{s} e^{-Ls}. \tag{1.6}$$

1.1.2 Supply Chain Management Processes

The supply chain [19, 109, 148] of a modern enterprise consists of suppliers, factories, warehouses, distribution centres, and retailers, which may reside at different locations in the world. In a supply chain, raw materials are processed, and products are made, which are then delivered to the customer at the right place at the right time in the right quantity. IPDT models can be employed effectively to describe supply chain management (SCM) processes [19, 21, 109, 144, 148].

For a factory, the production process can be depicted as shown in Figure 1.2, where a delay L is involved. The difference of the product completed and the customer demand d is stored in a buffer to generate the net stock S. The customer demand d is considered as an exogenous disturbance signal, and its effects on the stock level S must be dealt with properly by manipulating the production order rate u.

This model can also be used to describe the dynamics of warehouses or retailers, where the delay L is the transportation delay involved.

1.1.3 Communication Networks

Communication networks [124] have been among the fastest-growing areas in engineering in the last two decades. In most data communication networks, user data

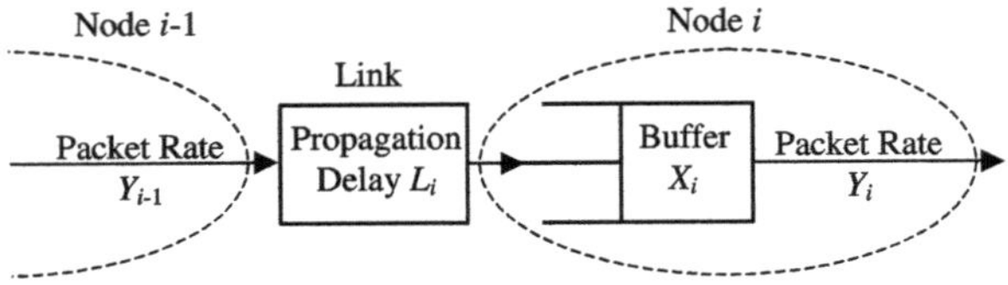

(a) sketch

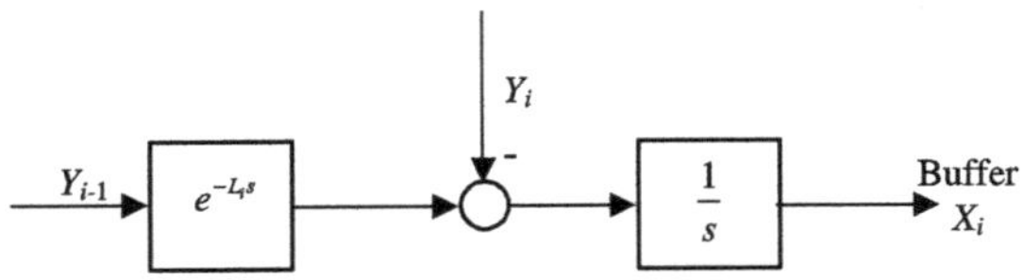

(b) its model in block diagrams

Fig. 1.3 A single-node connection in networks

is decomposed into a string of packets, including additional information added to ensure its safe arrival at the correct destination. The packets can be rather long or of variable length, *e.g.* as in the Internet, and small and of fixed length, *e.g.* as in Asynchronous Transfer Mode (ATM) networks. The data packets are stored in a buffer or queue at the source node and then sent to the buffer or queue at the next node when resources become available. This is repeated in the network until the packets arrive at the destination. Figure 1.3(a) depicts the process when packets are sent from Node $i-1$ to Node i. A propagation delay L_i is involved in the communication link. The mathematical model of such a single-node connection is shown in Figure 1.3(b), where the buffer is modelled as an integrator. Because of the delays involved, communication networks need to be well controlled to maintain network stability and quality-of-service (QoS); see for example [27, 44, 68, 105].

1.1.4 Other Examples

Other examples of IPDT processes include the high-pressure steam flowing to a steam turbine generator in a power plant [37], the heating of well-insulated batch systems, the batch preparation of solutions by addition of chemicals to solvent, and batch distillation processes where the mixture is added into the reboiler drum and a certain distillate composition is made to track a given trajectory by manipulating the internal reflux ratio [75]. The totally heat-integrated distillation columns [40] also exhibit a dynamics with an integrator.

1.2 Overview of the Book

The book is organised into two parts. In the first part, (one-degree-of-freedom) Proportional-Integral-Derivative (PID) based control schemes are addressed, and, in the second part, two-degree-of-freedom control schemes are addressed.

In Chapter 2, the use of PID controllers for integrating processes with dead time is introduced. In particular, the PID control techniques specialised for this kind of processes are analysed. After having introduced PID controllers, identification methods suitable to be applied in an industrial context are presented. Both open-loop and (relay-feedback-based) closed-loop techniques are considered. Then, tuning methodologies are explained. Without aiming at presenting all the methods proposed in the literature, different approaches are highlighted, with the purpose of showing how the problem can be tackled from different point of views. After having presented empirical tuning rules, analytical approaches are discussed, in particular those based on the Internal Model Control. Then, frequency-domain design techniques are explained. Eventually, techniques based on optimisation criteria are presented, focusing on the optimisation of integral and H_∞ criteria.

In Chapter 3, a procedure for the determination of the complete set of stabilising parameters of the PI controller for IPDT processes is presented together with the achievable stability margins. The stability regions under the PID control are also are addressed.

Chapter 4 deals with the performance assessment problem. In particular, a methodology for the assessment of the performance of a PID controller for IPDT processes is described, and a method for the retuning of the controller, if the performance is not satisfactory, is also explained.

In Chapter 5, the recently developed Plug&Control strategy is applied to integral processes with dead time. It will be shown that it represents a useful tool for the fast tuning of the controller at the start-up of the process.

In Chapter 6, after having presented the standard two-degree-of-freedom control scheme and a method for designing a two-state time-optimal feedforward controller, a method for the design of a noncausal feedforward action, based on input–output inversion, is applied to a PID-based feedback control scheme, and its implementation in an industrial context is addressed.

In Chapter 7, the design of a PID–PD controller is described. Typically, a PD controller is employed in an inner loop in the feedback path, while a PI controller is employed in the feedforward path. Similar approaches, namely, PID–PD and PID–I are also addressed.

In Chapter 8, Smith-predictor-based control schemes are presented. In the last few years, there have been many papers addressing the design of a modified Smith predictor in order to achieve high performance for an integral process with large dead time, in particular by ensuring zero steady-state error when a constant load disturbance occurs. This chapter describes some of these approaches by presenting the devised schemes, the tuning of the parameters, and the anti-windup strategies.

In Chapter 9, a control scheme which combines the advantages of the Smith predictor and the PID controller is presented. The controller is inherently a PID-type controller in which the integral action is implemented using a delay unit rather than a pure integrator. The setpoint response and the disturbance response are decoupled from each other and can be designed separately. The robustness of the scheme is also analysed.

In Chapter 10, a disturbance observer-based control scheme is explained. The controller can be designed to reject ramp disturbances and step disturbances, even arbitrary disturbances. The case to achieve deadbeat responses for step disturbances is also discussed in detail.

In Chapter 11, the performance achievable by the disturbance-observer-based control scheme is analysed quantitatively. In particular, four specifications—(normalised) maximal dynamic error, maximal decay rate, (normalised) control action bound, and approximate recovery time—are given to characterise the step-disturbance response.

Finally, in Chapter 12, some practical issues concerned with the controller implementation are discussed.

Part I
PID Control Schemes

Chapter 2
PID Control

The use of Proportional-Integral-Derivative (PID) controllers for integrating processes with dead time is discussed in this chapter. Since PID controllers are the most adopted controllers in industry and there are many different design methods, only those techniques specialised for IPDT processes are discussed in this chapter. After having introduced PID controllers, identification methods suitable to be applied in an industrial context are presented. Both open-loop and (relay-feedback-based) closed-loop techniques are considered. Then, tuning methods are explained. Without aiming at presenting all the tuning methods proposed in the literature, different approaches are highlighted with the purpose of showing how the problem can be tackled from different points of view. After having presented empirical tuning rules, analytical design techniques are explained, in particular those based on the Internal Model Control. Then, frequency domain approaches are discussed. Subsequently, techniques based on optimisation criteria are presented, in particular those based on the optimisation of integral criteria and on H_∞ loop shaping. Note that the classification done hereafter is subjective because, actually, there might be overlap between the different methods considered (for example, the Internal Model Control strategy is analytical, but, at the same time, it minimises integral criteria).

2.1 PID Controllers

2.1.1 Basic Principles

PID controllers are the most adopted controllers in industrial settings owing to their relative ease of use and the satisfactory performance they are capable to provide for the great majority of processes. Indeed, the cost/benefit ratio they can achieve is difficult for other kinds of controllers to compete with. Because of their widespread use, many techniques have been proposed for their design, namely, for the tuning of the parameters and for the implementation of additional functionalities that improve their performance; see for example [6, 83, 132].

A. Visioli, Q.-C. Zhong, *Control of Integral Processes with Dead Time*,
Advances in Industrial Control,
DOI 10.1007/978-0-85729-070-0_2, © Springer-Verlag London Limited 2011

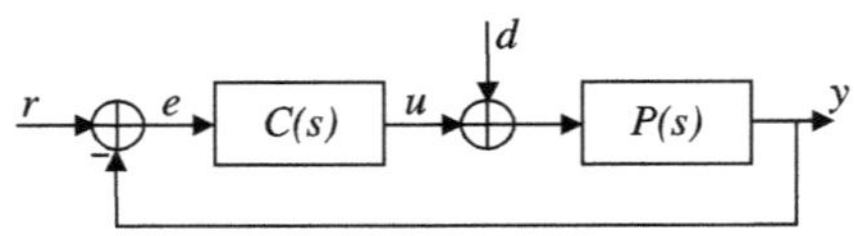

Fig. 2.1 Standard unity-feedback control scheme

A PID controller consists of the sum of three control actions, namely, a control action proportional to the control error, a control action proportional to the integral of the control error, and a control action proportional to the first derivative of the control error. Indeed, the proportional action implements the typical operation of increasing the control variable when the control error is large (with appropriate sign). The integral action is related to the past values of the control error and allows the reduction to zero of the steady-state error when a step reference signal is applied or a constant load disturbance d occurs. The derivative action is based on the predicted future values of the control error and has therefore a great potential in improving the control performance as it can anticipate an incorrect trend of the control error and counteract for it.

In its basic form, the control action can be expressed as

$$u(t) = K_p\left(e(t) + \frac{1}{T_i}\int_0^t e(v)\,dv + T_d\frac{d}{dt}e(t)\right), \tag{2.1}$$

where $e(t) = r(t) - y(t)$ is the control error (see Figure 2.1), K_p is the proportional gain, T_i is the integral time constant, and T_d is the derivative time constant. The corresponding transfer function is

$$C(s) = \frac{U(s)}{E(s)} = K_p\left(1 + \frac{1}{T_i s} + T_d s\right). \tag{2.2}$$

Actually, the transfer function (2.2) is not proper. In order to make it proper, a filter is usually added to the derivative term (note that this also reduces the detrimental effect of the high-frequency measurement noise), so that

$$C(s) = \frac{U(s)}{E(s)} = K_p\left(1 + \frac{1}{T_i s} + \frac{T_d s}{1 + \frac{T_d}{N}s}\right), \tag{2.3}$$

where the value of N generally assumes a value between 1 and 33, although in the majority of the practical cases, its setting falls between 8 and 16 (usually 10) [1]. Alternatively, the whole control action can be filtered, yielding the following transfer function:

$$C(s) = K_p\left(1 + \frac{1}{T_i s} + T_d s\right)\frac{1}{T_f s + 1}, \tag{2.4}$$

where the time constant T_f has to be selected so that the first-order filter does not influence the dynamics of the PID controller significantly and the high-frequency noise is filtered appropriately at the same time.

It is worth noting at this point that there are other possible configurations for a PID controller. In fact, in addition to Expression (2.2), which is called the *ideal* form (or *non-interacting* form), the three control actions can also be implemented in the so-called *series* or *interacting* form, namely (the filter is omitted for simplicity),

$$C(s) = K_p'\left(\frac{T_i's+1}{T_i's}\right)(T_d's+1), \tag{2.5}$$

or, alternatively, in *parallel* form as

$$C(s) = K_p + \frac{K_i}{s} + K_d s. \tag{2.6}$$

Translation formulae can be employed to determine the values of the parameters of an equivalent PID controller in a given form starting from the parameters of the PID controller in another form. However, it has to be stressed that the ideal form is more general than the series form because the controller can be designed with complex conjugate zeros.

2.1.2 Improvements

Modifications of the basic control law (2.1) are usually implemented to cope with practical issues. For example, the derivative action is often applied to the process output y instead of to the control error, so that an impulse in the control signal is avoided when a step signal is applied to the set-point. In this case, the derivative action $u_d(t)$ is expressed as

$$u_d(t) = -\frac{K_p}{T_d}\frac{d}{dt}y(t). \tag{2.7}$$

Thus, a general formula for the derivative action can be written as

$$u_d(t) = \frac{K_p}{T_d}\left(\gamma\frac{d}{dt}r(t) - \frac{d}{dt}y(t)\right), \tag{2.8}$$

where $\gamma = 1$ if the derivative action is applied to the control error and $\gamma = 0$ if the derivative action is applied to the process output.

Further, a set-point weight can be applied also to the proportional action in order to reduce the overshoot in the set-point step response (this is done at the expense of an increase of the rise time), so that the proportional action $u_p(t)$ is expressed as

$$u_p(t) = K_p\big(\beta r(t) - y(t)\big), \tag{2.9}$$

where the value of β is selected in the interval [0, 1]. The use of a set-point weight is particularly useful when specifications on both set-point following and load disturbance rejection tasks have to be addressed at the same time. Indeed, a fast load

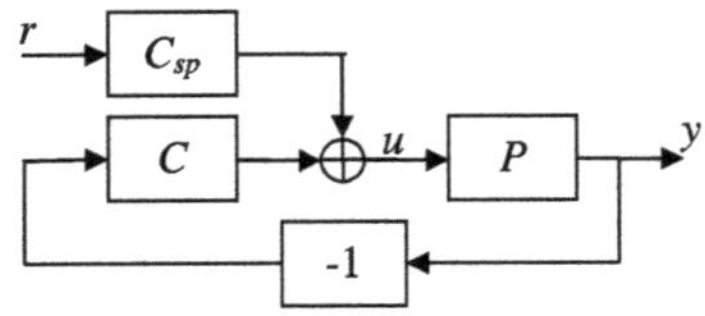

Fig. 2.2 Two-degree-of-freedom PID control scheme

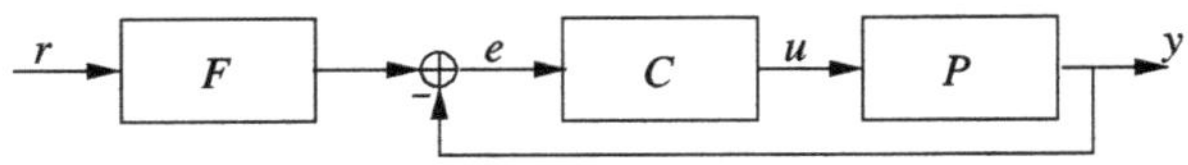

Fig. 2.3 Equivalent two-degree-of-freedom PID control scheme

disturbance rejection is achieved with a controller that provides a large bandwidth, which, in turn, gives an oscillatory set-point step response. By using a set-point weight, the control scheme represented in Figure 2.2 is actually implemented with

$$C(s) = K_p\left(1 + \frac{1}{T_i s} + T_d s\right) \tag{2.10}$$

and

$$C_{sp}(s) = K_p\left(\beta + \frac{1}{T_i s} + \gamma T_d s\right). \tag{2.11}$$

It appears that the load disturbance rejection does not depend on the weight β and therefore can be addressed separately from the set-point following task. Thus, the PID parameters can be selected to achieve a high load disturbance rejection performance, and then the set-point following performance can be recovered by suitably selecting the value of the parameter β. An equivalent control scheme is shown in Figure 2.3, where

$$F(s) = \frac{1 + \beta T_i s + \gamma T_i T_d s^2}{1 + T_i s + T_i T_d s^2}. \tag{2.12}$$

Here it is more apparent that the set-point weight is able to smooth the (step) set-point signal in order to damp the response to a set-point change.

If these modifications of the basic control law are considered, the general so-called *ISA form* (or *beta–gamma*) PID control law can be derived:

$$\begin{aligned} u(t) &= K_p\left(\beta r(t) - y(t) + \frac{1}{T_i}\int_0^t e(\tau)\,d\tau + T_d\left(\frac{d(\gamma r(t) - y_f(t))}{dt}\right)\right), \\ \frac{T_d}{N}\frac{dy_f(t)}{dt} &= y(t) - y_f(t), \end{aligned} \tag{2.13}$$

where $\beta \in [0, 1]$ and $\gamma \in \{0, 1\}$.

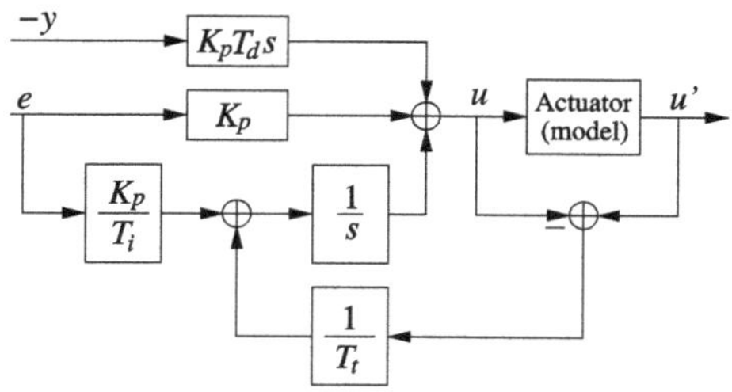

Fig. 2.4 The back-calculation anti-windup scheme

Finally, the integrator windup phenomenon has to be avoided. When a set-point change is applied, the control variable might reach and remain at the actuator saturation limit during the transient response. In this case the system operates as in the open-loop case because the actuator is at its maximum (or minimum) limit, regardless of the process output value. The control error decreases more slowly than in the ideal case (where there is no saturation limits), and therefore the integral term becomes large (it *winds up*). Thus, even when the value of the process variable attains that of the reference signal, the controller still saturates due to the integral term, and this generally leads to large overshoots and long settling time. In order to avoid this, an anti-windup strategy should be implemented. This can be done according to the so-called *conditional integration* technique, where the integral action is frozen when the actuator saturates and, at the same time, the control error and the control variable have the same sign, or according to the so-called *back-calculation* approach shown in Figure 2.4, where the integral action is reduced by a term proportional to the saturation level of the actuator. The parameter T_t, called the tracking time constant, determines the amount of reduction of the integral term.

2.2 Identification

System identification is a topic that has been and is extensively investigated, and many solutions have been proposed in the literature that can be applied in general to industrial processes. In the following sections, methodologies that have been specifically devised for the estimation of the parameters of a integral process are presented, by considering open-loop and closed-loop techniques that can be easily applied in practical cases. Simple continuous-time models are considered because these are commonly employed for the tuning of PID controllers.

2.2.1 Open-loop Identification

Open-loop identification techniques are based on the evaluation of the response of the process to particular signals. They have to be applied starting from an equilibrium point of the system.

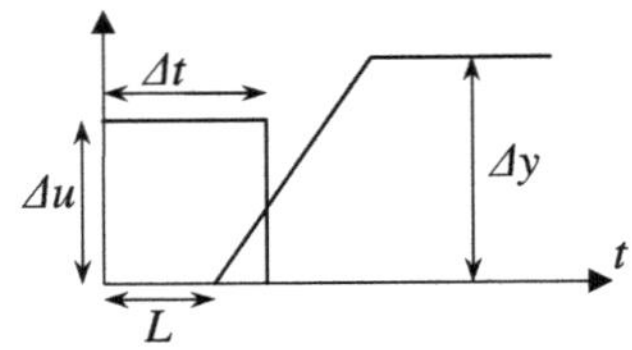

Fig. 2.5 Identification based on the impulse response

2.2.1.1 Using an Impulsive Input

In general, the most employed open-loop identification methods used for industrial system are based on the evaluation of a step response. However, in the case of integral processes, which are not asymptotically stable, the step response would tend to infinity, and this may not be acceptable in practice. An alternative sensible procedure is to apply an impulse to the system. Consider the IPDT system

$$P(s) = \frac{K}{s}e^{-Ls} \tag{2.14}$$

at a steady state (denote the output value as y_0) and apply an impulse of amplitude Δu and duration Δt as the input signal. This is shown in Figure 2.5 together with the corresponding output. The dead time L can be estimated by considering the time interval between the step transition and the time instant when the process output leaves its previous value, namely, $y > y_0$ (without loss of generality, it has been assumed that $\Delta u > 0$ and $K > 0$). It should be noted that in practice the measurement noise needs to be taken into consideration. A simple sensible solution is to define a noise band *NB* [8] (whose amplitude should be equal to the amplitude of the measurement noise) and to rewrite the condition as $y > y_0 + NB$. The value of K can then be derived, by considering that the value of the process output variation Δy is equal to the area of the impulse input multiplied by K, as

$$K = \frac{\Delta y}{\Delta u \Delta t}. \tag{2.15}$$

The estimation of the parameter K is based just on the steady-state value of the process output, and therefore it is easy to cope with measurement noise. Obviously, the estimation will be perfect when the process dynamics are represented perfectly by expression (2.14). This is not the case if the true process dynamics are different. For example, consider the process

$$P(s) = \frac{1}{s(s+1)}e^{-0.5s}. \tag{2.16}$$

If the open-loop impulse response (plotted as a solid line in Figure 2.6) is evaluated, the following IPDT model is obtained (note that the noise-free case is considered for the sake of simplicity):

$$P(s) = \frac{1}{s}e^{-0.5s}, \tag{2.17}$$

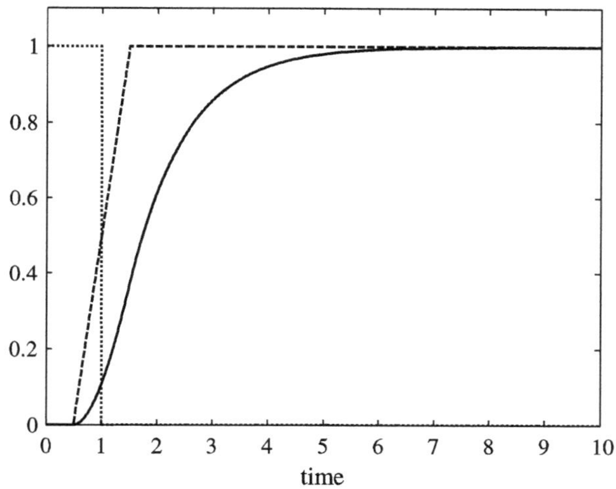

Fig. 2.6 Application of the identification method based on the impulse response. *Dotted line*: impulse input. *Solid line*: process response. *Dashed line*: response of the identified model

which gives the impulse response plotted as a dashed line in Figure 2.6. It turns out that the identification procedure has not captured the dynamics represented by the pole in $s = -1$, and the two responses are therefore somewhat different.

2.2.1.2 Using a Square Wave Input

The gain K can be identified via applying a square wave $u(t)$ with period P and amplitude Δu centred around a nominal value u^* [75]. Since the input function u is discontinuous at the time instants $t_d = P/2, P, 3P/2, \ldots$, the output response is continuous but not differentiable at these time instants. The gain K can be therefore computed as

$$K = \frac{\dot{y}(t_d^+) - \dot{y}(t_d^-)}{\Delta u}, \tag{2.18}$$

where $\dot{y}(t_d^+)$ and $\dot{y}(t_d^-)$ are respectively the time derivatives of the process output from the right and the left. The situation is shown in Figure 2.7, where the process $P(s) = 0.1/s$ is taken as an example (the dead time is omitted as it does not affect the identification of K). The main advantage of the method is that, by evaluating the gain at each discontinuity time instant, a time-varying gain can be estimated, and this fact can be exploited in the design of the controller (see [75] for an example related to a batch distillation column). However, it has to be taken into account that the differentiation procedure is very sensitive to the measurement noise and therefore data should be appropriately filtered before applying it.

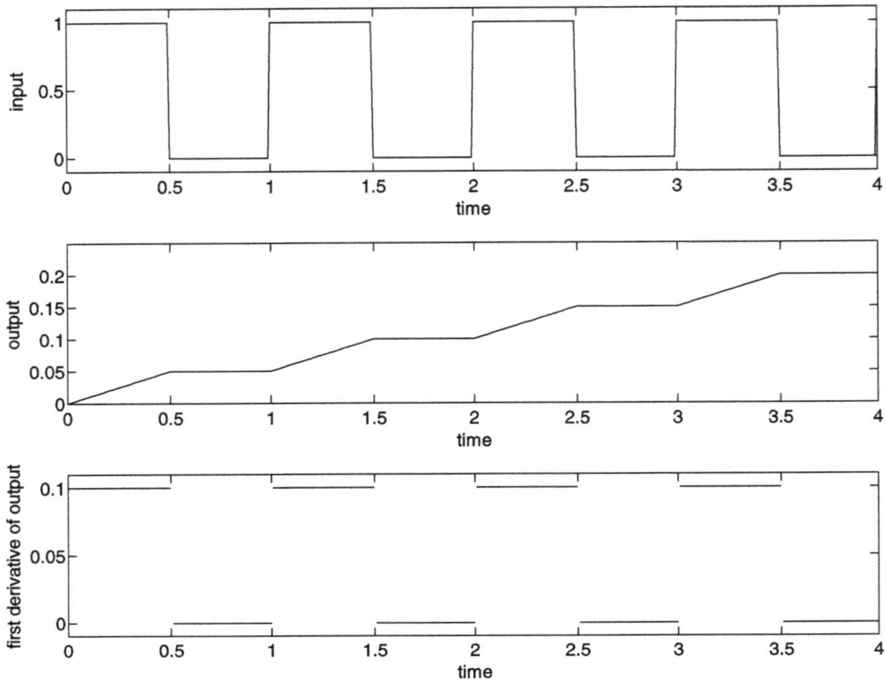

Fig. 2.7 Application of the identification method based on a square wave input

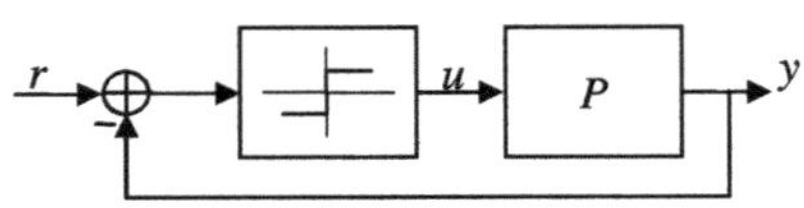

Fig. 2.8 Relay-feedback control scheme

2.2.2 Closed-loop Identification

Closed-loop identification techniques are usually based on the use of a relay feedback controller or, alternatively, on the evaluation of the response to a set-point change. Different methods in these contexts are presented hereafter.

2.2.2.1 Based on the Relay Feedback Methods

The closed-loop identification techniques employed for industrial processes are typically based on a relay-feedback experiment [151] (see Figure 2.8). The rationale of the use of the relay-feedback controller is to evaluate the obtained process output oscillation in order to obtain a nonparametric model of the process [3], namely its ultimate gain K_u and the ultimate frequency ω_u, in analogy with the original idea of the ultimate sensitivity experiment of Ziegler and Nichols [166], where the control system is led to the stability limit. Then, starting from these parameters, a transfer function of the process can be determined, if necessary.

Standard Relay-feedback Method The original relay-feedback experiment proposed in [3] involves the use of a standard symmetrical relay in order to generate a persistent oscillatory response of the process output. Denoting by h the amplitude of the relay and by A the amplitude of the output oscillations, the value of the ultimate gain can be calculated, by applying the describing function theory, as

$$K_u = \frac{4h}{\pi A}, \tag{2.19}$$

while the value of the ultimate period $P_u = 2\pi/\omega_u$ is simply the period of the obtained output oscillation. It appears that only the amplitude h of the relay has to be selected by the user. This should be done in order to provide an output oscillation of sufficient amplitude to be well distinguished from the measurement noise, but at the same time it should not be too high so that the process is perturbed as less as possible (and the normal production is not interrupted). Indeed, it is worth stressing that the estimation of the output oscillation is sensitive to the measurement noise and therefore some filtering techniques have to be applied [140]. In addition to the advantage of having just one parameter to be selected by the user and of being performed in closed-loop, so that the process is kept close to the set-point value, the main valuable feature of this identification technique is that the identification experiment can start also if the process is not at an equilibrium point. Further, possible load disturbances that might occur during the experiment can be detected easily by the change to asymmetric pulses in the control variable.

In any case, because of the adoption of the describing function theory, the obtained values of the ultimate gain and ultimate period are approximated. As an example, if the process

$$P(s) = \frac{1}{s} e^{-0.2s} \tag{2.20}$$

is considered, the result obtained by employing a relay-feedback controller is shown in Figure 2.9 (note that a set-point step equal to one has been applied at time $t = 0$). By noting that $h = 1$ and $A = 0.205$, the parameters obtained by applying the identification procedure are (see (2.19)) $K_u = 6.21$ and $P_u = 0.82$, while the true ones are $K_u = 7.85$ and $P_u = 0.80$. Actually, the slight difference is usually acceptable if the parameters are employed for the tuning of a PID controller.

The transfer function (2.14) can be determined from the estimated values of the ultimate gain and ultimate period by using the following expressions [26]:

$$L = 0.25 P_u, \tag{2.21}$$

$$K = \frac{4A}{h P_u}. \tag{2.22}$$

When these are applied to the previous example, then $L = 0.2$ and $K = 1$, which are equal to the true values.

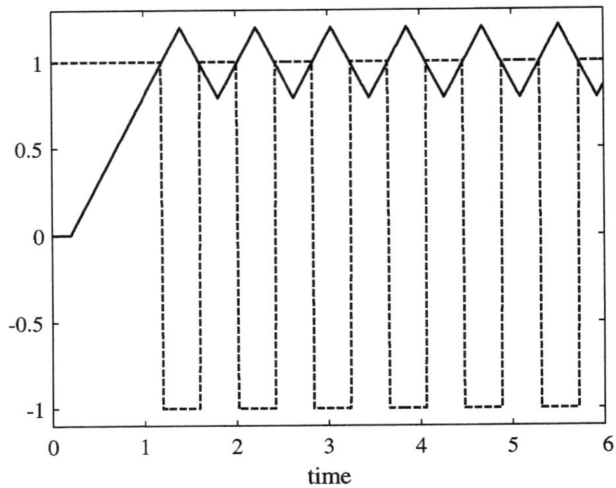

Fig. 2.9 Example of a relay-feedback identification experiment. *Dashed line*: control variable. *Solid line*: process variable

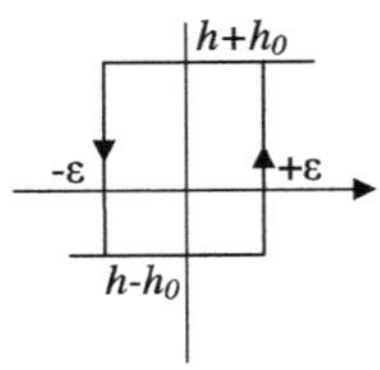

Fig. 2.10 Biased relay with hysteresis

Biased Relay with Hysteresis A biased relay with hysteresis (see Figure 2.10) can also be employed effectively for the estimation of a process transfer function (2.14) [59]. In fact, under an asymmetrical biased relay feedback test (see Figure 2.11), the output response of an IPDT system converges to a limit cycle described as

$$\begin{aligned} y_+(t) &= K(h+h_0)t + K(h-h_0)t_0, & t &\in \left[0, P^+\right], \\ y_-(t) &= K(h-h_0)(t+t_0) + K(h+h_0)P^+, & t &\in [0, P^-], \end{aligned} \tag{2.23}$$

where $y_+(t)$ denotes the monotonically ascending part for $t \in [0, P^-]$, corresponding to $t \in [P^+, P_u]$ in the limit cycle, while $y_-(t)$ denotes the monotonically descending part for $t \in [P^+, P_u]$, and $P_u = P^+ + P^-$ is the oscillation period. Based on these analytical expressions, the process parameters can be determined by evaluating an experiment with a biased relay with hysteresis feedback controller. In particular, the process dead time L can be determined as the time interval to attain the positive peak A^+ of the process output response from a relay switch point in a

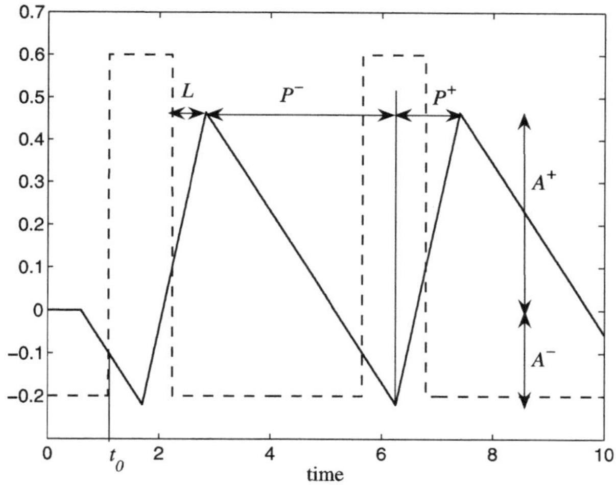

Fig. 2.11 Experiment with a biased relay with a hysteresis feedback controller

negative half period P^- of the relay, and the process gain K can be determined as

$$K = \frac{A^+ - A^-}{(h + h_0)P^+}. \tag{2.24}$$

The use of a biased relay is particularly useful if the system to be estimated is a second-order integral process plus dead time (SOIPDT) described by the transfer function

$$P(s) = \frac{K}{s(Ts+1)} e^{-Ls}. \tag{2.25}$$

In this case, the analytical expression of the limit cycle can be written as

$$\begin{aligned} y_+(t) &= K(h+h_0)(t-T) + K(h-h_0)t_0 + 2Kh_0TEe^{-t/T}, & t &\in \left[0, P^+\right], \\ y_-(t) &= K(h-h_0)(t+t_0-T) + K(h+h_0)P^+ + 2Kh_0TFe^{-t/T}, & t &\in (0, P^-], \end{aligned} \tag{2.26}$$

where

$$E = \frac{1 - e^{-P^-/T}}{1 - e^{-P_u/T}}$$

and

$$F = \frac{1 - e^{-P^+/T}}{1 - e^{-P_u/T}}.$$

Based on these expressions, the times to attain the extreme values of $y_+(t)$ and $y_-(t)$, denoted respectively as t_{P^+} and t_{P^-}, can be determined as

$$t_{P^+} = T \ln \frac{2h_0E}{h+h_0} \tag{2.27}$$

and

$$t_{P-} = T \ln \frac{2h_0 F}{h - h_0}. \tag{2.28}$$

Denote the time interval to reach the minimum of $y_+(t)$ from the initial relay switch point in a positive half period of the relay as t^*_{P+}; then

$$t_{P+} = t^*_{P+} - L, \tag{2.29}$$

and

$$t_{P-} = t^*_{P-} - L, \tag{2.30}$$

where t^*_{P-} is the time interval to reach the maximum of $y_-(t)$ from the initial relay switch point in a negative half period of the relay. By (2.29) and (2.30), we have

$$t_{P+} - t_{P-} = t^*_{P+} - t^*_{P-}. \tag{2.31}$$

Substitute (2.27) and (2.28) into (2.31); then

$$\ln \frac{h_0 - h}{h_0 + h} + \ln \frac{1 - e^{-P^-/T}}{1 - e^{-P_+/T}} = \frac{t^*_{P+} - t^*_{P-}}{T}. \tag{2.32}$$

Note that the process response at the oscillation frequency can be formulated as

$$P(j\omega_u) = \frac{\int_0^{P_u} y(t) e^{-j\omega_u t}\, dt}{\int_0^{P_u} u(t) e^{-j\omega_u t}\, dt} = A_u e^{j\varphi_u}. \tag{2.33}$$

Substitute the process model (2.25) into (2.33); then

$$\frac{K}{\omega_u \sqrt{T^2 \omega^2 + 1}} = A_u \tag{2.34}$$

and

$$-L\omega_u - \frac{\pi}{2} - \arctan(T\omega_u) = \varphi_u. \tag{2.35}$$

It can be easily derived that

$$K = A_u \omega_u \sqrt{T^2 \omega^2 + 1} \tag{2.36}$$

and

$$L = -\frac{1}{\omega_u} \left[\varphi_u + \frac{\pi}{2} + \arctan(T\omega_u) \right]. \tag{2.37}$$

In case $L/T > 1$, $y_+(t)$ can decrease monotonically for $t \in [0, P^+]$, while $y^-(t)$ can increase monotonically for $t \in [0, P^-]$. Thus, there exists

$$t^*_{P+} = t^*_{P-} = L. \tag{2.38}$$

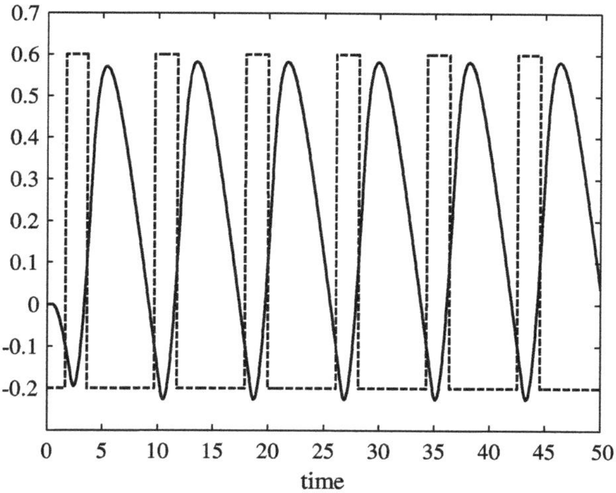

Fig. 2.12 Experiment for the estimation of a SOIPDT process

In this case, the process time constant can be derived from (2.37) as

$$T = \frac{1}{\omega_u} \tan\left(-L\omega_u - \frac{\pi}{2} - \varphi_u\right), \tag{2.39}$$

and the process gain can be determined straightforward from (2.36). However, the value of T cannot be determined from (2.39) if

$$\tan\left(-L\omega_u - \frac{\pi}{2} - \varphi_u\right) < 0. \tag{2.40}$$

The algorithm for the identification of the SOIPDT process (2.25) can be summarised as follows [59]:

1. Measure P^+, P^-, t^*_{P+}, and t^*_{P-} from the limit cycle.
2. Compute $P(j\omega_u)$ from (2.33).
3. Compute the process dead time from (2.38) and check whether (2.40) is satisfied. If yes, go to step 6.
4. Compute the process time constant T from (2.39) and check whether $L/T < 1$ is satisfied. If yes, go to step 6.
5. Determine the process gain K from (2.36). If both (2.40) and $L/T < 1$ are not satisfied, then terminate.
6. Determine T from (2.32) by applying the Newton–Raphson iteration method. Set $T = t^*_{P+}$(or $T = t^*_{P-}$) as the initial estimation of T.
7. Determine the process gain K from (2.36).
8. Determine the process dead time L from (2.37).

As an example of this algorithm, the process (2.16) is considered. The result of the application of the biased relay feedback control scheme is shown in Figure 2.12. The following parameters are obtained: $t^*_{p+} = 0.8$, $t^*_{p-} = 1.76$, $P^+ = 2.04$, $P^- = 6.14$, $P_u = 8.18$, (namely, $\omega_u = 0.77$), $A_u = 1.04$. By applying the algorithm, the process parameters are estimated correctly.

An alternative approach for the estimation of the parameters of a SOIPDT process has been proposed in [35]. The method is based on the application of the relay feedback controller to a first-order-plus-dead-time (FOPDT) process described by the following transfer function:

$$P(s) = \frac{K}{Ts+1}e^{-Ls}. \tag{2.41}$$

In this case, the process parameters can be estimated by applying the following formulae, which are based again on the describing function analysis:

$$K = \frac{a_0}{u_0}, \tag{2.42}$$

$$T = \frac{P_u}{2\pi \frac{A^+ - A^-}{2}} \sqrt{K^2(\alpha^2 + \beta^2) - \left(\frac{A^+ - A^-}{2}\right)^2}, \tag{2.43}$$

$$L = \frac{P_u}{2\pi}\left[\pi - \tan^{-1}\left(\frac{2\pi T}{P_u}\right) + \tan^{-1}\left(\frac{\alpha}{\beta}\right)\right], \tag{2.44}$$

where

$$\alpha = \frac{-4h_0\varepsilon}{\pi \frac{A^+ - A^-}{2}}, \tag{2.45}$$

$$\beta = \frac{h}{\pi \frac{A^+ - A^-}{2}}\left(\sqrt{\left(\frac{A^+ - A^-}{2}\right)^2 - (a_0 - \varepsilon)^2} + \sqrt{\left(\frac{A^+ - A^-}{2}\right)^2 - (a_0 + \varepsilon)^2}\right), \tag{2.46}$$

and a_0 and u_0 are the values of the DC components of the oscillations at the process output and input. In case the process is described by a SOIPDT transfer function, in principle it is sufficient to differentiate the output of the relay and then apply the preceding formulae (2.42)–(2.44). Obviously, differentiating the relay output gives impulses at the zero crossings that are not acceptable in practical cases because of the actuator constraints. In order to cope with this problem, the ideal impulses can be substituted by pulses with finite amplitude and short pulse width. Actually, this introduces an approximation that might significantly affect the estimation result.

It is worth stressing that, when a relay feedback controller is employed, some filtering techniques should be applied because the estimation of the output oscillation is sensitive to the measurement noise. Furthermore, the use of an asymmetrical relay represents a sort of disturbance to the process since it causes the operating point to drift.

2.2.2.2 Based on the Closed-loop Step Responses

A closed-loop identification method which is an alternative to the use of a relay feedback controller consists of evaluating the set-point step response of the IPDT process (2.14) with simple proportional controller $C(s) = K_p$ [115]. In this case, the closed-loop transfer function is (see Figure 2.1)

$$\frac{Y(s)}{R(s)} = \frac{K'e^{-Ls}}{s + K'e^{-Ls}}, \tag{2.47}$$

where

$$K' = K_p K. \tag{2.48}$$

By using a first-order Padè approximation, namely $e^{-Ls} \cong (1-0.5Ls)/(1+0.5Ls)$, Expression (2.47) can be rewritten as

$$\frac{Y(s)}{R(s)} = \frac{K'(1+0.5Ls)e^{-Ls}}{0.5\frac{L}{K'}s^2 + (\frac{1}{K'} - 0.5L)s + 1} \tag{2.49}$$

or, equivalently,

$$\frac{Y(s)}{R(s)} = \frac{K'(1+0.5Ls)e^{-Ls}}{\tau_e^2 s^2 + 2\tau_e \zeta s + 1} \tag{2.50}$$

with

$$\tau_e = \sqrt{\frac{L}{2K'}} \tag{2.51}$$

and

$$\zeta = \sqrt{\frac{K'}{2L}}\left(\frac{1}{K'} - \frac{L}{2}\right). \tag{2.52}$$

The effective time constant τ_e and the damping coefficient ζ of the closed-loop system can be estimated by considering the closed-loop response parameters y_{M1}, y_{m1}, y_{M2}, y_∞ (namely, the first peak value, the first minimum value, the second peak value, and the steady-state value) and Δt as shown in Figure 2.13. In particular, the following formulae can be employed [26]:

$$\zeta = -\frac{\ln \frac{y_{M2}-y_\infty}{y_{M1}-y_\infty}}{\sqrt{4\pi^2 + \left(\ln \frac{y_{M2}-y_\infty}{y_{M1}-y_\infty}\right)^2}}, \tag{2.53}$$

or, alternatively,

$$\zeta = -\frac{\ln \frac{y_\infty-y_{m1}}{y_{M1}-y_\infty}}{\sqrt{\pi^2 + \left(\ln \frac{y_\infty-y_{m1}}{y_{M1}-y_\infty}\right)^2}}, \tag{2.54}$$

and

$$\tau_e = \frac{\Delta t}{\pi}\sqrt{1-\zeta^2}. \tag{2.55}$$

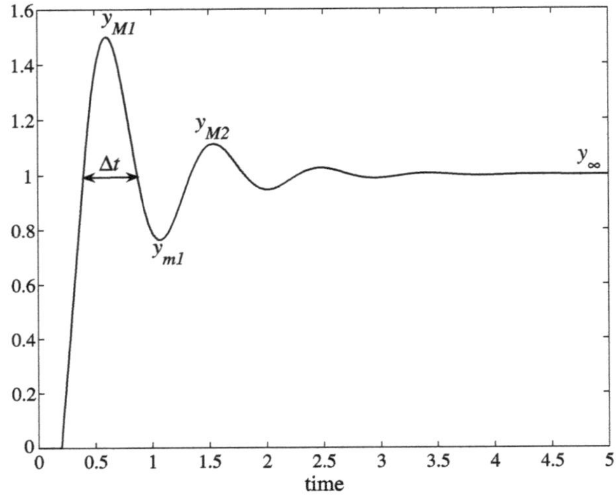

Fig. 2.13 Experiment for the estimation of a SOIPDT process

The values of L and K' (and consequently of K) can be eventually determined, by solving Equations (2.51) and (2.52), as

$$L = \frac{2.828\Delta t}{\pi}\sqrt{0.5+\zeta^2-\sqrt{\zeta^2(1+\zeta^2)}}\sqrt{1-\zeta^2}, \tag{2.56}$$

$$K = \frac{1.414\pi}{\Delta t\sqrt{1-\zeta^2}}\sqrt{0.5+\zeta^2-\sqrt{\zeta^2(1+\zeta^2)}}. \tag{2.57}$$

It is worth noting that the value of the proportional gain of the controller has to be sufficiently high to provide an oscillatory set-point step response.

As an illustrative example of application of the identification procedure, consider process (2.20). If a proportional feedback controller $K_p = 5$ is employed, the resulting set-point step response is shown in Figure 2.13 with $y_{M1} = 1.5$, $y_{m1} = 0.76$, $y_{M2} = 1.11$, and $y_\infty = 1$. By applying either (2.53) or (2.54) we have $\zeta = 0.23$, and, as a consequence, from (2.56), (2.57), and (2.48) the resulting values of the process parameters are $K = 1.09$ and $L = 0.23$. The slight discrepancy between the true and estimated values is due to the Padè approximation.

2.3 Tuning Methods

A large number of tuning methods have been proposed in literature over the last seventy years. They are based on different approaches and aim at solving different control problems. In fact, as it has already been mentioned in Section 2.1, there are different (possibly conflicting) control tasks that have to be addressed in practical cases. In particular, set-point following and/or load disturbance rejection are usually of main concern. In general, a good load disturbance rejection performance is achieved with a high-gain controller, which gives an oscillatory set-point step response on the other side. If both specifications have to be considered, the problem can be solved by employing a set-point weight. It is worth stressing at this point that, although the use of a set-point weight yields a two-degree-of-freedom control system, its use is addressed in this part of the book because the two controllers (C and F) in Figure 2.3 are actually strongly correlated. In Part II of the book, in a context different from PID controllers, the two controllers are designed separately in a two-degree-of-freedom scheme.

In addition to the performance in the set-point following and/or load disturbance rejection tasks, the performance in the filtering of measurement noise, the robustness of the control scheme and the control effort have to be considered as well in the selection of the PID parameters. For this reason, it is difficult to make the best choice among the many tuning rules that are available.

In the following subsections, instead of providing a comprehensive review of all the tuning rules that are available for integral processes (see [83] for this purpose), different approaches with the aim of showing the peculiarities of integral processes are highlighted.

2.3.1 Empirical Formulae

Empirical formulae for the tuning of the PID controllers have been devised soon after PID controllers appeared in the industry at the beginning of the last century. The most well-known formulae are those devised by Ziegler and Nichols in the 1940s [166]. There are two kinds of formulae that are based respectively on the parametric model (2.14) and on the nonparametric model given by the ultimate gain K_u and the ultimate frequency ω_u. They are shown in Tables 2.1 and 2.2, respectively. It is worth noting that the Ziegler–Nichols tuning rules aim at providing a good load disturbance rejection performance (in particular, a quarter decay ratio in the load disturbance step response), and this implies that the damping ratio of the closed-loop system is in general too low to achieve a satisfactory set-point following performance (namely, the step response is too oscillatory with a big overshoot). As already mentioned, this issue can be addressed by employing a set-point weight for the proportional action. As an example, consider the process

$$P(s) = \frac{0.0506}{s} e^{-6s}. \tag{2.58}$$

By applying the PID controller tuning rule shown in Table 2.1, the parameters are $K_p = 3.95$, $T_i = 12$, and $T_d = 3$. The resulting set-point and load disturbance rejection step response is shown in Figure 2.14 as a solid line. If a set-point weight $\beta = 0.4$ is employed, the result obtained is the one shown with a dashed line. If Table 2.2 is used to determine the PID parameters (note that $K_u = 5.19$ and $P_u = 24$), the parameters are $K_p = 3.11$, $T_i = 12$, and $T_d = 3$. The corresponding results (obtained again with and without the set-point weight) are plotted in Figure 2.15. As it can be easily expected by evaluating the controller parameters, the method based

Table 2.1 Ziegler–Nichols tuning rules based on a parametric model

Controller	K_p	T_i	T_d
P	$\frac{1}{KL}$	–	–
PI	$\frac{0.9}{KL}$	$3L$	–
PID	$\frac{1.2}{KL}$	$2L$	$\frac{L}{2}$

Table 2.2 Ziegler–Nichols tuning rules based on a non-parametric model

Controller	K_p	T_i	T_d
P	$0.5K_u$	–	–
PI	$0.4K_u$	$0.8P_u$	–
PID	$0.6K_u$	$0.5P_u$	$0.125P_u$

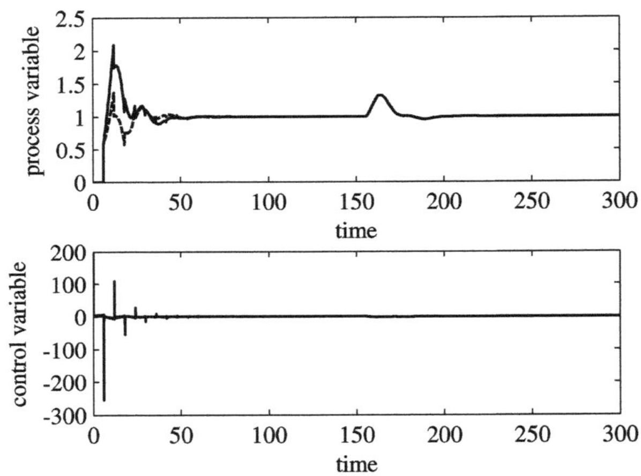

Fig. 2.14 Results obtained with the Ziegler–Nichols tuning rules based on a parametric model of the process. *Solid line*: PID controller with no set-point weight. *Dashed line*: PID controller with set-point weight

on the nonparametric model of the process provides a less oscillatory response. In any case, the use of a set-point weight is actually essential in reducing the excessive overshoot that occurs because of the aggressive tuning conceived to achieve a fast load disturbance rejection.

2.3.2 Analytical Methods

In contrast to empirical tuning rules, analytical methods are based on the determination of the controller parameters by exploiting explicitly the expressions that involve the transfer function of the closed-loop system.

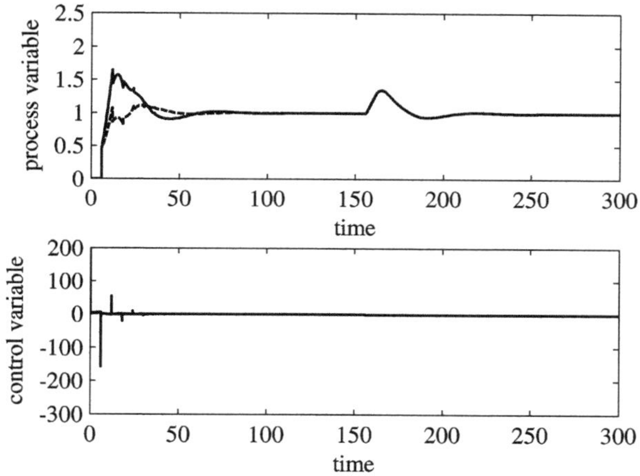

Fig. 2.15 Results obtained with the Ziegler–Nichols tuning rules based on a nonparametric model of the process. *Solid line*: PID controller with no set-point weight. *Dashed line*: PID controller with set-point weight

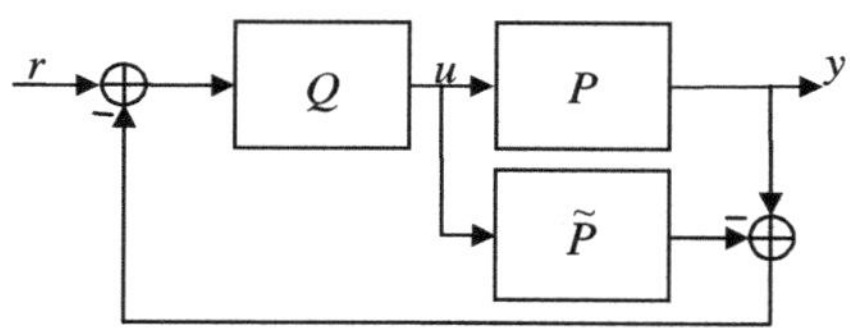

Fig. 2.16 The general Internal Model Control scheme

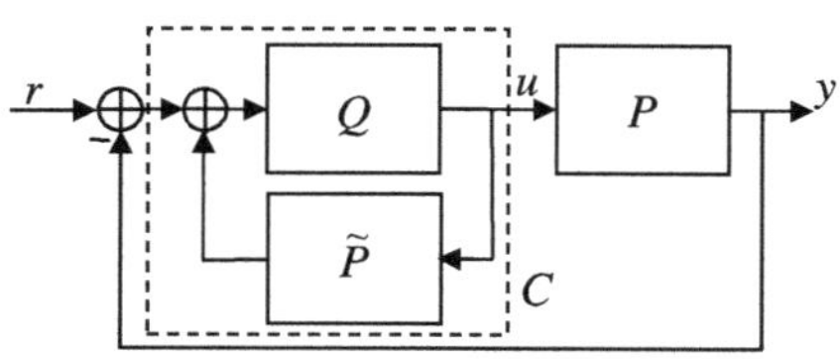

Fig. 2.17 The equivalent unity-feedback control scheme

2.3.2.1 Internal Model Control

Internal Model Control (IMC) [76] is a well-known control design approach where the trade-off between nominal performance and robustness is explicitly addressed. This is obtained by including a model of the process in the controller implementation, according to the scheme of Figure 2.16, where $\tilde{P}$ denotes the model of the process P, and the controller Q determines the value of the control variable u. Note that this scheme is equivalent to the unity-feedback scheme of Figure 2.1 by simply selecting (see Figure 2.17)

$$C(s) = \frac{Q(s)}{1 - \tilde{P}(s)Q(s)}. \tag{2.59}$$

The design of the controller is in general performed by considering $Q(s) = \tilde{Q}(s)F(s)$ and by selecting $\tilde{Q}(s)$ in order to achieve an optimal performance for a given input disregarding the model uncertainties (namely, by considering $\tilde{P}(s) = P(s)$) and input constraints. Then, $F(s)$ is selected as a low-pass filter of an appropriate order in order to achieve robust stability and robust performance. In particular, $\tilde{Q}(s)$ can be determined to minimise the integrated square error (ISE)

$$\int_0^{\infty} e^2(t)\,dt. \tag{2.60}$$

This is obtained by factoring the model $\tilde{P}(s)$ into two parts:

$$\tilde{P}(s) = \tilde{P}_+(s)\tilde{P}_-(s), \tag{2.61}$$

where $\tilde{P}_+(s)$ is the all-pass portion of the transfer function ($\tilde{P}_+(0) = 1$) including all the RHP zeros and delays of $\tilde{P}(s)$, having the form

$$\tilde{P}_+(s) = e^{-Ls}\prod_i \frac{-\alpha_i s + 1}{\alpha_i s + 1}, \tag{2.62}$$

where α_i^{-1} are all the RHP zeros, and L is the dead time.

Alternatively, if it is desired to minimise the integrated absolute error (IAE)

$$\int_0^{\infty} \left|e(t)\right| dt, \tag{2.63}$$

$\tilde{P}_+(s)$ in the factorisation (2.61) should be chosen as

$$\tilde{P}_+(s) = e^{-Ls}\prod_i (-\alpha_i s + 1). \tag{2.64}$$

Then, $\tilde{Q}(s) = \tilde{P}_-^{-1}(s)$. At this point, $F(s)$ has to be selected in order to have a proper controller $Q(s)$. With the inclusion of the filter, the transfer function of the closed-loop system of Figure 2.16 becomes $P(s)\tilde{Q}(s)F(s)$ (again, by considering $\tilde{P}(s) = P(s)$). Thus, in order to achieve a null steady-state error in the presence of a step set-point signal, it has to be $F(0) = 1$. In this context, a natural choice is to select

$$F(s) = \frac{1}{(\lambda s + 1)^n}, \tag{2.65}$$

where the order n is such that the controller $Q(s)$ is proper. In order to have a null steady-state error when a ramp signal is applied to the set-point, in addition to the requirement $F(0) = 1$, the following expressions needs to be satisfied as well:

$$\frac{d}{ds}\big(\tilde{P}(s)Q(s)\big)\bigg|_{s=0} = 0$$

Table 2.3 IMC-based PID tuning rules

Approximation	$K_p K$	T_i	T_d	T_f	Comments
$e^{-Ls} \cong 1 - Ls$	$\frac{2(L+\lambda)}{2L^2+4\lambda L+\lambda^2}$	$2(L+\lambda)$	–	$\frac{L\lambda^2}{2L^2+4\lambda L+\lambda^2}$	ISE optimal
$e^{-Ls} \cong 1 - Ls$	$\frac{L+2\lambda}{(L+\lambda)^2}$	$L+2\lambda$	–	–	IAE optimal
$e^{-Ls} \cong \frac{1-\frac{L}{2}s}{1+\frac{L}{2}s}$	$\frac{3L+2\lambda}{2L^2+4\lambda L+\lambda^2}$	$3L+2\lambda$	$\frac{2L(L+\lambda)}{3L+2\lambda}$	$\frac{L\lambda^2}{2L^2+4\lambda L+\lambda^2}$	ISE optimal
$e^{-Ls} \cong \frac{1-\frac{L}{2}s}{1+\frac{L}{2}s}$	$\frac{2}{L+\lambda}$	$2(L+\lambda)$	$\frac{L(L+2\lambda)}{2(L+\lambda)}$	–	IAE optimal

or, equivalently,

$$\frac{d}{ds}\left(\tilde{P}_+(s)F(s)\right)\Big|_{s=0} = 0.$$

A possible solution is

$$F(s) = \frac{(2\lambda - \tilde{P}'_+(0))s + 1}{(\lambda s + 1)^2}. \tag{2.66}$$

By restricting the analysis to IPDT processes (2.14), it has to be stressed that, in order to have a null steady-state error with a constant load disturbance, it is necessary for the controller to have a pole at the origin. This corresponds to having a null steady-state error when a ramp signal is applied to the set-point, and therefore the filter transfer function (2.66) has to be considered. Then, if the dead time is approximated as

$$e^{-Ls} \cong 1 - Ls \tag{2.67}$$

and the IMC design procedure is applied, $C(s)$ becomes a PI controller with or without an output filter, respectively, depending on whether the factorisation (2.62) or (2.64) is applied. If the dead time is approximated as

$$e^{-Ls} \cong \frac{1 - \frac{L}{2}s}{1 + \frac{L}{2}s}, \tag{2.68}$$

then $C(s)$ becomes a PI controller with or without an output filter, respectively, depending on whether the factorisation (2.62) or (2.64) is applied. Thus, the application of the IMC design naturally yields the tuning rules shown in Table 2.3 [108], where the only parameter to be selected by the user is λ, which handles the trade-off between aggressiveness and robustness (and control activity). Indeed, increasing the value of λ implies that the closed-loop time constant increases and the robustness of the control system to plant/model mismatch increases. Conversely, decreasing the value of λ implies that the speed of response increases but the system is less robust. Different practical recommendations for the choice of λ have been proposed in the literature. For example, the advice $\lambda > L/4$ or $\lambda > L/2$ is given in [108] corresponding to the factorisation form (2.62) or (2.64), while the suggestion $\lambda = L\sqrt{10}$ is given in [2]. It is worth stressing that if the PID controller has a form

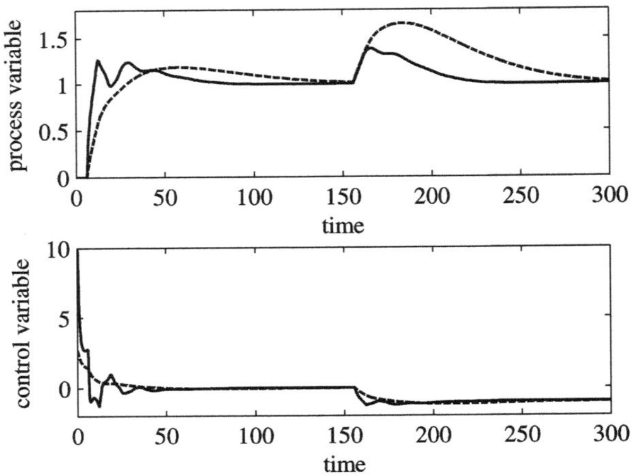

Fig. 2.18 Results obtained with the IMC-based tuning rules. *Solid line*: $\lambda = 6$. *Dashed line*: $\lambda = 18.97$

different from the output filtered ideal one (2.4), simple conversion formulae can be employed [15].

As an illustrative example, consider again process (2.58). By applying the tuning rule that yields a PID controller with an output filter and by initially selecting $\lambda = L = 6$, the parameters are $K_p = 2.35$, $T_i = 30$, $T_d = 4.8$, and $T_f = 0.86$. The response to a set-point and load disturbance step signals is shown in Figure 2.18 as a solid line. Conversely, by selecting $\lambda = L\sqrt{10} = 18.97$, it is $K_p = 1.25$, $T_i = 55.95$, $T_d = 5.36$, and $T_f = 2.43$. The corresponding set-point and load disturbance step responses are plotted again in Figure 2.18 as a dashed line. As expected, a bigger value of λ yields a less aggressive and more robust control system, namely, the overshoot in the set-point step response is reduced, the rise time is increased, and the control effort is reduced as well. Conversely, a more sluggish load disturbance response is obtained.

2.3.2.2 Matching the Coefficients of the Closed-loop Transfer Function

If just the set-point following performance is addressed, a simple method to tune the PID controller is to match the coefficients of the numerator and denominator polynomial of the closed-loop transfer function [14]. If an IPDT process transfer function (2.14) is considered and a PID controller (2.2) is employed, the closed-loop transfer function from the set-point r to the output y is

$$\frac{Y(s)}{R(s)} = \frac{(K_1 q + K_2 + K_3 q^2)e^{-q}}{q^2 + (K_1 q + K_2 + K_3 q^2)e^{-q}}, \tag{2.69}$$

where

$$q = Ls, \tag{2.70}$$

$$K_1 = K_p K L, \tag{2.71}$$

$$K_2 = \frac{K_1 L}{T_i}, \tag{2.72}$$

$$K_3 = K_1 \frac{T_d}{L}. \tag{2.73}$$

By using the first-order Padè approximation for the exponential term at the denominator, Expression (2.69) becomes

$$\frac{Y(s)}{R(s)} = \frac{(K_1 q + K_2 + K_3 q^2)(1+0.5q)e^{-q}}{(1+0.5q)q^2 + (K_1 q + K_2 + K_3 q^2)(1-0.5q)}. \tag{2.74}$$

By imposing that this closed-loop transfer function is equal to one, it results

$$K_1 = 1, \tag{2.75}$$

$$K_2 = 0, \tag{2.76}$$

$$K_3 = 0.5, \tag{2.77}$$

that is (see (2.71)–(2.73)),

$$K_p = \frac{1}{KL}, \tag{2.78}$$

$$T_i = \infty, \tag{2.79}$$

$$T_d = 0.5L. \tag{2.80}$$

Indeed, a PD controller results. This is in accordance with the intuition that a pole at the origin of the complex plane to ensure a null steady-state error is already present in the process transfer function. In order to employ the integral action in any case (for example to cope with possible load disturbances), it is sufficient to impose that the closed-loop transfer function (2.74) is equal to $\alpha > 1$ (note that if $\alpha = 1$, the same tuning rules as before are obtained). This is reasonable because the steady-state error will be zero in any case (because of the presence of the integrator) and because with a PI or PID controller an overshoot occurs in any case in the set-point step response. Thus, by considering α as a tuning parameter, the following expressions are obtained:

$$(1-\alpha)K_1 + 0.5(1+\alpha)K_2 = 0, \tag{2.81}$$

$$0.5(1+\alpha)K_1 + (1-\alpha)K_3 = \alpha, \tag{2.82}$$

$$(1+\alpha)K_3 = \alpha, \tag{2.83}$$

which yields

$$K_p = \frac{1}{KL}\frac{4\alpha^2}{(1+\alpha^2)}, \tag{2.84}$$

$$T_i = 0.5L\left(\frac{1+\alpha}{\alpha-1}\right), \tag{2.85}$$

$$T_d = 0.25L\left(\frac{1+\alpha}{\alpha}\right). \tag{2.86}$$

If a PI controller is considered, by following the same reasoning as before the following rules are obtained:

$$K_p = \frac{1}{KL}\frac{2\alpha}{1+\alpha}, \tag{2.87}$$

$$T_i = 0.5L\left(\frac{1+\alpha}{\alpha-1}\right). \tag{2.88}$$

It appears that all the PID parameters depend on the value of α which has to be selected conveniently. In [14] it is suggested to choose $\alpha = 1.25$, namely, for the PID controller:

$$K_p = \frac{1.2346}{KL}, \tag{2.89}$$

$$T_i = 4.5L, \tag{2.90}$$

$$T_d = 0.45L. \tag{2.91}$$

As an illustrative example of the method, consider the process (2.58). By applying (2.89)–(2.91) the following parameters are determined: $K_p = 4.067$, $T_i = 27$, and $T_d = 2.7$. The resulting set-point and load disturbance unit step responses are shown as a solid line in Figure 2.19. They are compared with the results obtained by using a PI controller with $K_p = 3.66$ and $T_i = 27$ (see (2.87)–(2.88)), shown as a dashed line, and with the results obtained by using a PD controller with $K_p = 3.29$ and $T_d = 3$ (see (2.78)–(2.80)), shown as a dotted line. In the cases where the derivative action has been employed, a first-order filter has been applied, but its time constant has been selected so that its dynamics are actually negligible (note that the derivative filter has not been considered explicitly in the derivation of the tuning rules). This explains the large spikes in the control signal (no saturation of the actuator has been considered in order to avoid biasing the result). Further, as expected, the use of the PD controller provides a better set-point step response but exhibits a steady-state error in the presence of a constant load disturbance. The PID controller performs better than the PI controller, but a large overshoot appears in all of the cases.

2.3.2.3 Direct-synthesis-based Design

With respect to the method described in Section 2.3.2.2, an increase in the performance can be expected if a filtered PID controller (possibly with set-point weight) is employed. In this context, the method based on direct synthesis proposed in [107]

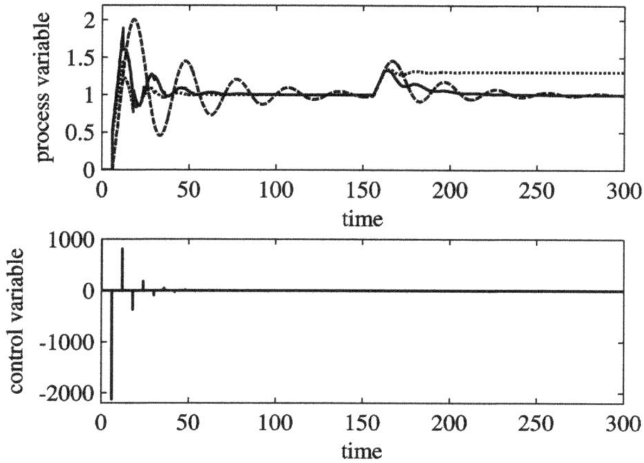

Fig. 2.19 Results obtained with the method based on matching the coefficients of the closed-loop transfer function. *Solid line*: PID controller. *Dashed line*: PI controller. *Dotted line*: PD controller

can be applied. In fact, if the scheme of Figure 2.1 is considered and if an appropriate desired closed-loop transfer function is selected, then the controller transfer function determined analytically has the required structure. Indeed, if the IPDT process (2.14) is considered and the desired closed-loop transfer function is selected as

$$\left(\frac{Y(s)}{R(s)}\right)_d = \frac{(\eta s+1)e^{-Ls}}{(\lambda s+1)^2}, \tag{2.92}$$

then, the corresponding controller transfer function can be determined as

$$C(s) = \frac{1}{P(s)}\frac{\left(\frac{Y(s)}{R(s)}\right)_d}{1-\left(\frac{Y(s)}{R(s)}\right)_d} = \frac{s}{K}\frac{(\eta s+1)}{[(\lambda s+1)^2-(\eta s+1)e^{-Ls}]}. \tag{2.93}$$

By applying the first-order Padè approximation $e^{-Ls} \cong (1-\frac{L}{2}s)/(1+\frac{L}{2}s)$ and by selecting $\eta = 2\lambda + L$, from (2.93) an output-filtered PID controller (2.4) is obtained, where

$$K_p = \frac{2\lambda + 1.5L}{K(\lambda^2 + 2\lambda L + 0.5L^2)}, \tag{2.94}$$

$$T_i = 2\lambda + 1.5L, \tag{2.95}$$

$$T_f = \frac{0.5\lambda^2 L}{\lambda^2 + 2\lambda L + 0.5L^2}. \tag{2.96}$$

The same method can be applied also to SOIPDT processes (2.25). In this case, the desired closed-loop transfer function has to be selected as

$$\left(\frac{Y(s)}{R(s)}\right)_d = \frac{(\eta_2 s^2 + \eta_1 s + 1)e^{-Ls}}{(\lambda s+1)^3}, \tag{2.97}$$

so that the controller is obtained as

$$C(s) = \frac{1}{P(s)} \frac{\left(\frac{Y(s)}{R(s)}\right)_d}{1-\left(\frac{Y(s)}{R(s)}\right)_d} = \frac{s(Ts+1)}{K} \frac{(\eta_2 s^2 + \eta_1 s + 1)}{[(\lambda s+1)^3 - (\eta_2 s^2 + \eta_1 s + 1)e^{-Ls}]}. \tag{2.98}$$

By using again a first-order Padè approximation, the controller transfer function becomes a PID controller in series with a lead/lag compensator:

$$C(s) = K_p\left(1 + \frac{1}{T_i s} + T_d s\right)\frac{(as+1)}{(bs+1)}, \tag{2.99}$$

where

$$K_p = \frac{\eta_1}{K(3\lambda^2 + 1.5\lambda L + 0.5\eta_1 L - \eta_2)}, \tag{2.100}$$

$$T_i = \eta_1, \tag{2.101}$$

$$T_d = \frac{\eta_2}{\eta_1}, \tag{2.102}$$

$$T_f = \frac{0.5\lambda^2 L}{\lambda^2 + 2\lambda L + 0.5L^2}, \tag{2.103}$$

$$a = 0.5L, \tag{2.104}$$

$$b = \frac{0.5\lambda^3 L}{T(3\lambda^2 + 1.5\lambda L + 0.5L\eta_1 - \eta_2)}, \tag{2.105}$$

$$\eta_1 = 3\lambda + L, \tag{2.106}$$

$$\eta_2 = \frac{(0.5L - T)\lambda^3 + (3T^2 - 1.5LT)\lambda^2 + 3LT^2\lambda + 0.5L^2T^2}{T(0.5L+T)}. \tag{2.107}$$

It appears that for both IPDT and SOIPDT processes, there is just one tuning parameter λ which handles the trade-off between aggressiveness and robustness. In [107] it is suggested to select λ in the range $[0.8L, 3L]$. Furthermore, it is suggested to use the set-point weight β in the range of 0.3–0.4 to reduce the overshoot in the step response.

As an illustrative example, by considering again process (2.58), the following parameters are determined, by applying (2.94)–(2.96) with $\lambda = L = 6$: $K_p = 3.29$, $T_i = 21$, $T_d = 2.57$, $T_f = 0.86$. The set-point weight has been fixed to 0.3. The resulting set-point and load disturbance unit step responses are shown in Figure 2.20.

2.3.3 *Frequency-domain Methods*

Tuning methods can also be developed by considering the frequency response of the system. Some examples are presented in the following subsections.

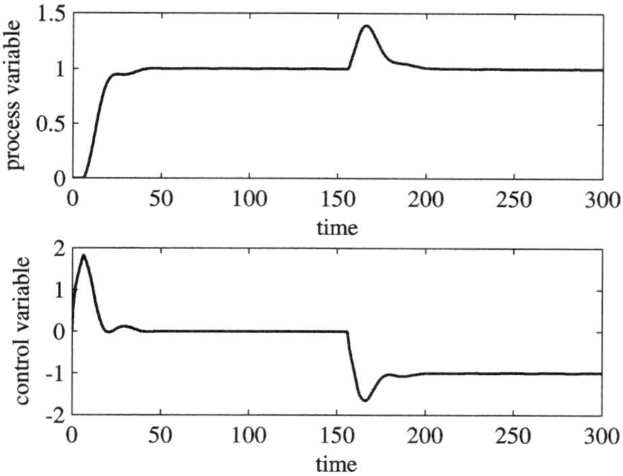

Fig. 2.20 Results obtained with the direct-synthesis-based design method

2.3.3.1 Based on the Maximum Peak-resonance Specification

The method proposed in [101] is based on a specification of a maximum peak resonance and is derived from the analysis of the Nichols chart of the series of the controller and of the process. In fact, with an integral process, the open-loop frequency response presents a phase maximum (see Figure 2.21). In this context, the controller parameters can be selected such that this maximum is located on the right-most point of the ellipse corresponding to the selected maximum peak resonance. Thus, the method can handle at the same time the maximum peak overshoot and the minimum phase and gain margins.

By considering a SOIPDT process (2.25) and a PI controller

$$C(s) = K_p\left(1 + \frac{1}{T_i s}\right), \tag{2.108}$$

specifying that the phase maximum (achieved at the frequency $\omega_{\max}$) of the open-loop frequency response $L(s) = C(s)P(s)$ is located at the right most point $(A_{\max}, \phi_{\max})$ of the contour corresponding to the desired maximum peak resonance M_r, yields the following system of three equations and three unknowns $(K_p, T_i, \omega_{\max})$:

$$\left.\frac{\partial \arg L(j\omega)}{\partial \omega}\right|_{\omega=\omega_{\max}} = 0, \tag{2.109}$$

$$\arg L(j\omega_{\max}) = \phi_{\max}, \tag{2.110}$$

$$\left|L(j\omega_{\max})\right| = A_{\max}. \tag{2.111}$$

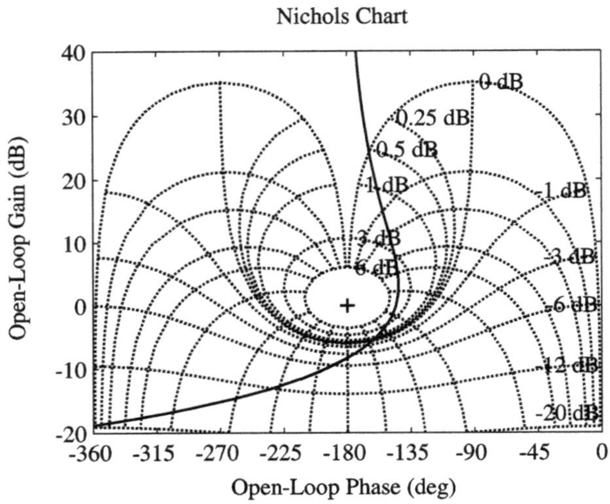

Fig. 2.21 Example of a Nichols chart for an integral process in series with a PI controller

A simple expression of the solution of this system can be obtained by approximating the $\arctan(x)$ as

$$\arctan(x) \cong \begin{cases} \frac{\pi}{2} - \frac{1}{x} & \text{if } x > 1, \\ x & \text{if } x \leq 1. \end{cases} \tag{2.112}$$

In this way the following expressions are obtained:

$$T_i = \frac{16(T+L)}{(2\phi_{\max}+\pi)^2}, \tag{2.113}$$

$$K_p = \frac{T_i A_{\max}}{K}\left[\frac{T_i^2\omega_{\max}^6 + \omega_{\max}^4}{T_i^2\omega_{\max}^2+1}\right]^{\frac{1}{2}}, \tag{2.114}$$

$$\omega_{\max} = \left[\frac{1}{T_i(T+L)}\right]^{\frac{1}{2}}. \tag{2.115}$$

In order to achieve a good compromise between the set-point following and the load disturbance rejection performance, it is suggested to select $M_r = 5$ dB (which corresponds to $A_{\max} = 1.21$ and $\phi_{\max} = -2.55$ rad). If a PID controller is employed, it has to be selected with transfer function

$$C(s) = K_p\left(\frac{T_i s + 1}{T_i s}\right)\frac{T_d s + 1}{T_f s + 1}, \tag{2.116}$$

so that, by selecting $T_d = T$ (namely, by applying a pole-zero cancellation), the previous case is obtained, where the time constant of the open-loop system is given by T_f.

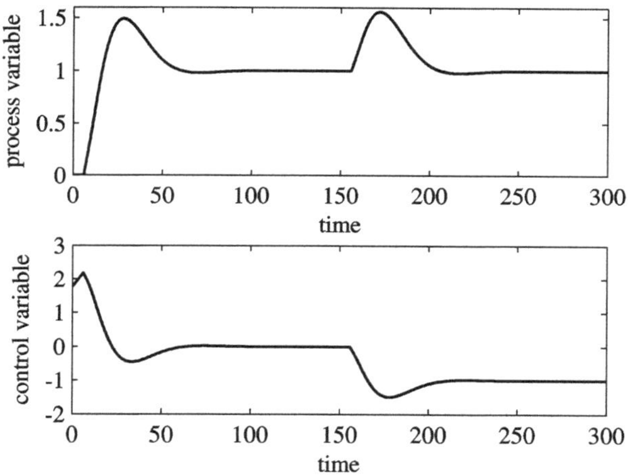

Fig. 2.22 Results obtained with the direct synthesis-based design method based on maximum peak-resonance specification

A particular interesting case is that given by IPDT processes, where $T = 0$. The tuning rules (2.113)–(2.115) are simplified significantly, and they can be related directly (possibly in an automatic tuning context) to the ultimate gain K_u and ultimate frequency P_u of the process as [102]

$$K_p = 0.34K_u, \tag{2.117}$$

$$T_i = 1.04P_u. \tag{2.118}$$

The same process (2.58) is employed as an illustrative example. By applying the proposed method, a PI controller with $K_p = 1.75$ and $T_i = 25$ ($\omega_{\max} = 0.082$) is determined. The resulting set-point and load disturbance unit step responses are plotted in Figure 2.22, while the Nichols chart is shown in Figure 2.21. Obviously, the overshoot in the set-point step response can be reduced by employing a set-point weight.

2.3.3.2 Based on the Minimisation of the Maximum Resonance Peak Value

An approach similar to that of Section 2.3.3.1 has been (previously) proposed in [122]. The approach starts by considering the fact that decreasing the integral time constant in a PI controller for an IPDT process implies that the stability margin of the system decreases as well. Thus, there is a minimum value of the integral time constant below which a reasonable damping cannot be achieved for a given system. The design method consists therefore in specifying the maximum resonance peak value and then in determining the smallest integral time constant for which this value is attained. If an IPDT process with a PI controller is considered, the system of equations (2.109)–(2.110) can be solved analytically without approximating the arctan function with expression (2.112). Thus,

$$\omega_{\max} = \frac{1}{T_i}\left[\frac{T_i - L}{L}\right]^{\frac{1}{2}}, \tag{2.119}$$

and, as a consequence,

$$\arg L(j\omega_{\max}) = -\pi - \frac{L}{T_i}\left(\frac{T_i}{L} - 1\right)^{\frac{1}{2}} + \arctan\left(\frac{T_i}{L} - 1\right)^{\frac{1}{2}}. \tag{2.120}$$

In [122] it is suggested to select $M_r = 2\,\text{dB}$, which corresponds to $\phi_{\max} = -2.23$ rad. With this value, by solving $\arg L(j\omega_{\max}) = \phi_{\max}$, it results

$$\frac{T_i}{L} = 8.75, \tag{2.121}$$

from which the value of the integral time constant T_i is determined. The proportional gain can be selected at this point in order for the resulting resonance peak value to be at a minimum. By applying this procedure to several numerical cases it has been found that the value of K_p that provides this result can be expressed as a function of the dead time and of the gain of the process:

$$K_p = \frac{0.487}{KL}. \tag{2.122}$$

If process (2.58) is considered, by applying the tuning rules (2.121)–(2.122), we have $T_i = 52.5$ and $K_p = 1.6$. The resulting set-point and load disturbance unit step responses are plotted in Figure 2.23.

2.3.3.3 Based on the Specification of the Desired Control Signal

An original approach, based on the specification of the desired control signal, has been proposed in [139]. Basically, for an IPDT process (2.14), the technique consists in selecting the transfer function between the set-point r and the control variable u as

$$Q(s) := \frac{U(s)}{R(s)} = \frac{s}{K}\frac{(2\xi\tau + L)s + 1}{\tau^2 s^2 + 2\xi\tau s + 1}, \tag{2.123}$$

where the time constant τ is chosen as

$$\tau = \alpha L. \tag{2.124}$$

It has to be stressed at this point that if a step signal of amplitude A_r is applied to the set-point, transfer function (2.123) implies that the desired control signal has an

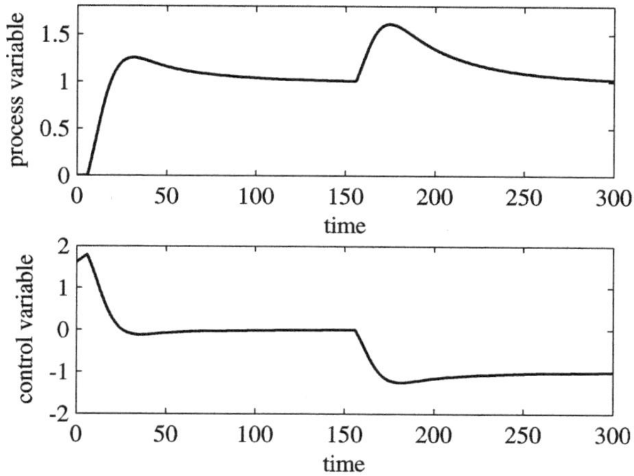

Fig. 2.23 Results obtained with the method based on the minimisation of the maximum resonance peak value

initial change of $(2\alpha\xi + 1)A_r/(\alpha^2 KL)$ and then decays exponentially to zero following a second-order system response with normalised time constant α and damping factor ξ. Given a desired damping factor (which is suggested to be either 0.707 or 1), this fact can be obviously exploited for the selection of the design parameter α in order to address the actuator constraints.

Given $Q(s)$, the desired closed-loop transfer function between the set-point r and the output y becomes $H(s) := Q(s)P(s)$, and the corresponding desired open-loop transfer function becomes

$$W(s) = \frac{H(s)}{1 - H(s)} = \frac{(2\alpha\xi + 1)Ls + 1}{\alpha^2 L^2 s^2 + 2\alpha\xi Ls + 1 - [(2\alpha\xi + 1)Ls + 1]e^{-Ls}} e^{-Ls}. \tag{2.125}$$

By considering now the PID controller transfer function (2.2), which can be rewritten as

$$C(s) = \frac{c_2 s^2 + c_1 s + c_0}{s}, \tag{2.126}$$

where

$$K_p = c_1, \qquad T_i = c_1/c_0, \qquad T_d = c_2/c_1, \tag{2.127}$$

the actual open-loop transfer function is given by

$$C(s)P(s) = \frac{c_2 s^2 + c_1 s + c_0}{s} \frac{K}{s} e^{-Ls}. \tag{2.128}$$

With the aim of matching $G(s)$ with the actual open-loop transfer function (2.128), the following transfer function can be defined ($s = j\omega$):

$$M(j\omega) = \frac{j\omega H(j\omega)}{P(j\omega)}. \tag{2.129}$$

This implies that the frequency domain error between $L(j\omega)$ and the actual open-loop transfer function is zero if

$$M(j\omega) = c_2(j\omega)^2 + c_1(j\omega) + c_0. \tag{2.130}$$

If two straight lines are employed to fit the real part $M_R(j\omega)$ of $M(j\omega)$ against ω^2 and the imaginary part $M_I(j\omega)$ of $M(j\omega)$ against ω through two frequencies ω_1 and ω_2, then the coefficients c_0, c_1, and c_2 can be determined analytically (note that, once the parameters α and ξ are selected, Expression (2.129) is known). The two frequencies can be conveniently selected as $\omega_1 = 2\pi/T_s$ and $\omega_2 = 2\omega_1$, where T_s is the desired closed-loop settling time, chosen as $(6\alpha + 1)L$ [138]. The solution is

$$c_0 = \frac{M_R(\omega_1) - M_R(\omega_2)}{3} + M_R(\omega_1), \tag{2.131}$$

$$c_1 = \frac{M_I(\omega_1)}{\omega_1}, \tag{2.132}$$

$$c_2 = \frac{M_R(\omega_1) - M_R(\omega_2)}{3\omega_1^2}, \tag{2.133}$$

from which the PID parameters can be easily derived from (2.127). At this point, it is worth considering the scaled Laplace transform $\hat{s} = sL$ (which naturally leads to a scaling in the time domain with a normalised time variable $\hat{t} = t/L$) and normalised frequencies $\hat{\omega}_1 = \omega_1 L$ and $\hat{\omega}_2 = 2\hat{\omega}_1$. The corresponding transfer function $M(j\hat{\omega})$ is therefore independent of the process parameters, and therefore the (scaled) PID parameters depend only on ξ and α. By considering the selected values of $\xi = 0.707$ and $\xi = 1$ and by interpolating the results for different values of α, the following tuning rules can be derived (the PID parameters are then conveniently rescaled):

$\xi = 0.707$:

$$K_p = \frac{1}{KL}\frac{1}{0.7138\alpha + 0.3904}, \tag{2.134}$$

$$T_i = L(1.4020\alpha + 1.2076), \tag{2.135}$$

$$T_d = \frac{1}{KL}\frac{1}{1.4167\alpha + 1.6999}; \tag{2.136}$$

$\xi = 1$:

$$K_p = \frac{1}{KL}\frac{1}{0.5080\alpha + 0.6208}, \tag{2.137}$$

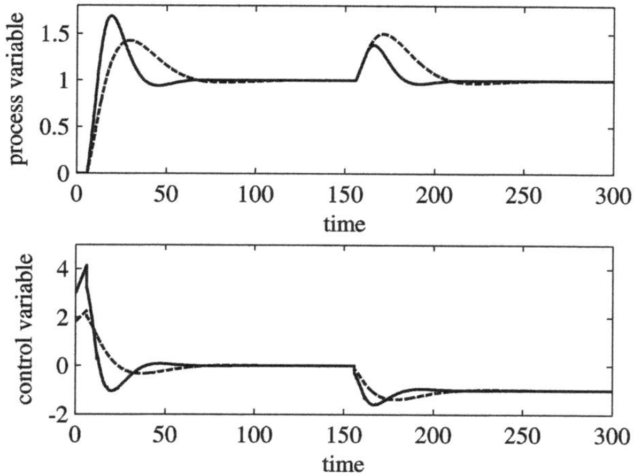

Fig. 2.24 Results obtained with the method based on the specification of the desired control signal (tuning rules (2.134)–(2.136)). *Solid line*: $\alpha = 1$. *Dashed line*: $\alpha = 2$

$$T_i = L(1.9885\alpha + 1.2235), \quad (2.138)$$

$$T_d = \frac{1}{KL}\frac{1}{1.0043\alpha + 1.8194}. \quad (2.139)$$

If a PI controller has to be employed, it is sufficient to substitute $T_d = 0$ in the above tuning rules.

The same process (2.58) is considered as an illustrative example. By applying the tuning rules (2.134)–(2.136) with $\alpha = 1$, the parameters $K_p = 2.98$, $T_i = 15.66$, and $T_d = 1.93$ are obtained, while with $\alpha = 2$, we obtain $K_p = 1.81$, $T_i = 24.07$, and $T_d = 1.32$. The results related to the set-point and load disturbance step response are shown in Figure 2.24. Conversely, if the tuning rules (2.137)–(2.139) are considered, we obtain $K_p = 2.92$, $T_i = 19.27$, and $T_d = 2.12$ for $\alpha = 1$ and $K_p = 2.01$, $T_i = 31.20$, and $T_d = 1.57$ for $\alpha = 2$. The corresponding results are shown in Figure 2.25. It can be seen that the parameter α can handle effectively the trade-off between aggressiveness and control effort.

2.3.4 Optimisation-based Methods

Tuning rules can be also obtained by minimising a suitable objective function. Methods developed in this context are explained hereafter.

2.3.4.1 Minimisation of the Integral Criteria

Significant attention has been paid by researchers in order to find the tuning of a PID controller that minimises integral performance criteria. This is, in fact, a way to

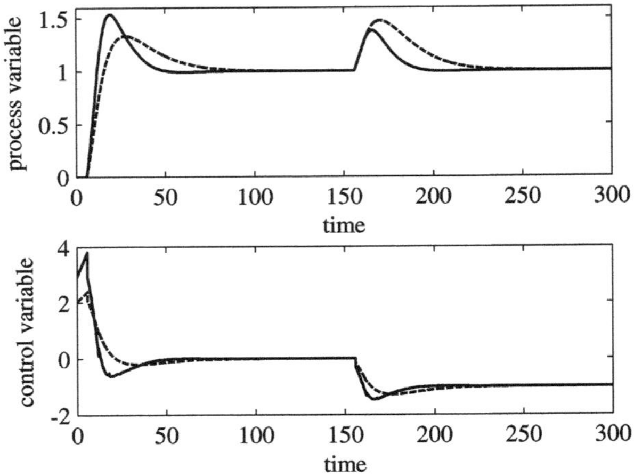

Fig. 2.25 Results obtained with the method based on the specification of the desired control signal (tuning rules (2.137)–(2.139)). *Solid line*: $\alpha = 1$. *Dashed line*: $\alpha = 2$

consider, at the same time, different control specifications, such as a small overshoot and a short settling time. In general, time-moment weighted integral performance indexes are considered. They are defined as

$$J_n(\theta) = \int_0^\infty t^n \big[e(t;\theta)\big]^2 dt, \quad n = 0, 1, 2, \tag{2.140}$$

where $\theta = [K_p, T_i, T_d]$ is the vector of (PID) parameters to be selected to minimise (2.140), and $e(t)$ is the system error. Note that $J_0(\theta)$ is denoted as the ISE (Integrated Square Error) criterion, while $J_1(\theta)$ and $J_2(\theta)$ are known respectively as the ITSE and ISTE criteria. A methodology for the determination of tuning formulae which relate the ideal PID coefficients (see (2.2)) to the process parameters K and L (see (2.14)) in order to minimise the objective functions (2.140) has been proposed in [128]. To this purpose, genetic algorithms [74], which are known to provide a global optimum of a problem in a stochastic framework, have been employed. Specifically, many simulations have been performed for different values of the parameter L (obviously, a different value of K results in a simple scaling of the proportional gain) and for different optimisation problems, *i.e.*, considering step changes both in the set-point and in the load disturbance and minimising the three adopted integral criteria (2.140). The optimal PID coefficients found by the genetic algorithms [74] in the different cases have then been analytically interpolated in order to derive suitable tuning rules. These are reported in Table 2.4 for the optimal set-point response and in Table 2.5 for the optimal load disturbance rejection. The symbol '–' which appears in Table 2.4 means that no integral action is required for that case, which is intuitive since the presence of an integrator in the plant assures by itself a zero steady-state error for set-point step changes and adding another integrator in the open-loop transfer function makes the achievement of an acceptable robustness more difficult. From these results it appears that increasing the value of

Table 2.4 PID tuning rules for optimal set-point response

	ISE	ITSE	ISTE
K_p	$\frac{1.03}{KL}$	$\frac{0.96}{KL}$	$\frac{0.90}{KL}$
T_i	–	–	–
T_d	$0.49L$	$0.45L$	$0.45L$

Table 2.5 PID tuning rules for optimal load disturbance response

	ISE	ITSE	ISTE
K_p	$\frac{1.37}{KL}$	$\frac{1.36}{KL}$	$\frac{1.34}{KL}$
T_i	$1.49L$	$1.66L$	$1.83L$
T_d	$0.59L$	$0.53L$	$0.49L$

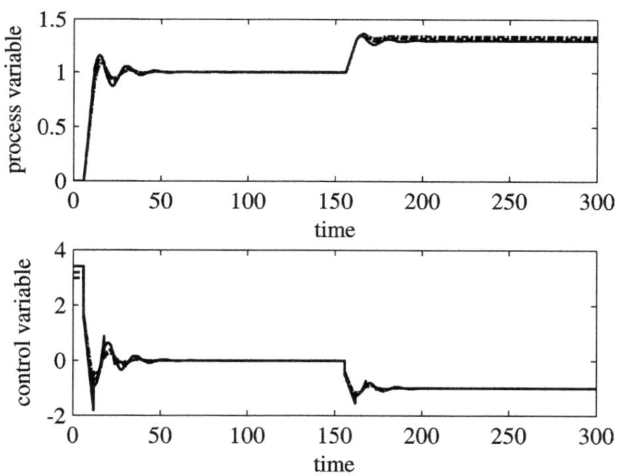

Fig. 2.26 Results obtained with the method based on the optimisation of integral performance indexes for set-point following task. *Solid line*: ISE. *Dashed line*: ITSE. *Dash-dot line*: ISTE

n from 0 to 2 in the performance index (2.140) implies that the PID gains have to be decreased.

As an illustrative example, if the set-point following task for process (2.58) is considered, by applying the tuning rule of Table 2.5, $K_p = 3.39$ and $T_d = 2.94$ are obtained for the ISE performance index, $K_p = 3.16$ and $T_d = 2.70$ are obtained for the ITSE performance index, and $K_p = 2.96$ and $T_d = 2.94$ are obtained for the ISTE performance index. The results related to both the set-point following and load disturbance rejection task are shown in Figure 2.26 (unit step signals are applied in both cases). It appears that, as expected, a steady-state error emerges in the presence of a constant load disturbance because there is no integral action in the controller. Conversely, if the load disturbance task is considered, by applying the

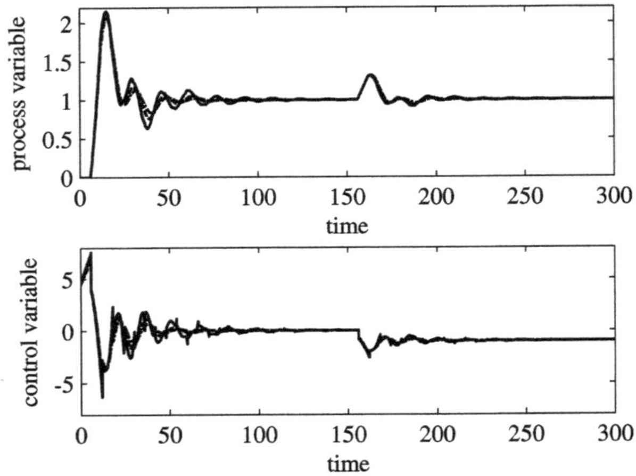

Fig. 2.27 Results obtained with the method based on the optimisation of integral performance indexes for load disturbance rejection task. *Solid line*: ISE. *Dashed line*: ITSE. *Dash-dot line*: ISTE

tuning rule of Table 2.5, $K_p = 4.51$, $T_i = 8.94$, and $T_d = 3.54$ are obtained for the ISE performance index, $K_p = 4.48$, $T_i = 9.96$, and $T_d = 3.18$ are obtained for the ITSE performance index, and $K_p = 4.41$, $T_i = 10.98$, and $T_d = 2.94$ are obtained for the ISTE performance index. The results again relating to both the set-point following and load disturbance rejection tasks are shown in Figure 2.27. As expected, the controller designed for the load disturbance rejection is more aggressive than the controller designed for the set-point following task.

2.3.4.2 Minimisation of an H_∞ Performance Index

A tuning methodology based on the optimisation of an H_∞ criterion has been proposed in [154]. By considering the Internal Model Control scheme of Figure 2.16 and the associated standard unity-feedback control scheme of Figure 2.1 where $C(s) = Q(s)/(1 - P(s)Q(s))$ (see also Figure 2.17), when perfect modelling is assumed, the sensitivity transfer function is

$$S(s) = \frac{1}{1 + C(s)P(s)} = 1 - P(s)Q(s), \tag{2.141}$$

and the complementary sensitivity transfer function is

$$H(s) = \frac{C(s)P(s)}{1 + C(s)P(s)} = P(s)Q(s). \tag{2.142}$$

The transfer function matrix $M(s)$ from the reference input r and the load disturbance input d to y and u is therefore

$$M(s) = \begin{bmatrix} H(s) & P(s)S(s) \\ Q(s) & -H(s) \end{bmatrix}. \tag{2.143}$$

The closed-loop system is internally stable if all the transfer functions in $M(s)$ are stable, which implies that $Q(s)$ is stable and satisfies the following constraints:

$$\lim_{s\to 0} S(s) = \lim_{s\to 0}\big[1 - P(s)Q(s)\big] = 0, \tag{2.144}$$

$$\lim_{s\to 0} \frac{d}{ds} S(s) = \lim_{s\to 0} \frac{d}{ds}\big[1 - P(s)Q(s)\big] = 0. \tag{2.145}$$

Then, a SOIPDT process (2.25), where the dead time is approximated by a first-order Taylor series, is considered:

$$P(s) = \frac{K(1 - Ls)}{s(Ts + 1)}. \tag{2.146}$$

The optimal performance criterion to be minimised by the control system is selected as

$$\big\| \Gamma(s)S(s) \big\|_{\infty}, \tag{2.147}$$

where $\Gamma(s)$ is a weighting function selected as

$$\Gamma(s) = \frac{1}{s}, \tag{2.148}$$

which implies that the closed-loop system input is a step signal.

If $\tilde{Q}(s) = Q(s)$ (namely, the filter $F(s)$ is neglected), minimising (2.147) yields

$$\Gamma(s)\big(1 - P(s)\tilde{Q}(s)\big) = L, \tag{2.149}$$

and therefore the optimal $\tilde{Q}(s)$ is determined as

$$\tilde{Q}(s) = \frac{s(Ts + 1)}{K}. \tag{2.150}$$

It appears that, in order to make $Q(s)$ proper, however, a filter $F(s)$ has to be employed.

If $T = 0$ (namely, an IPDT process is considered), it can be verified that a first-order filter does not satisfy the asymptotic tracking requirement. Thus, the filter is selected as the second-order transfer function

$$F(s) = \frac{as + 1}{(\lambda s + 1)^2}, \tag{2.151}$$

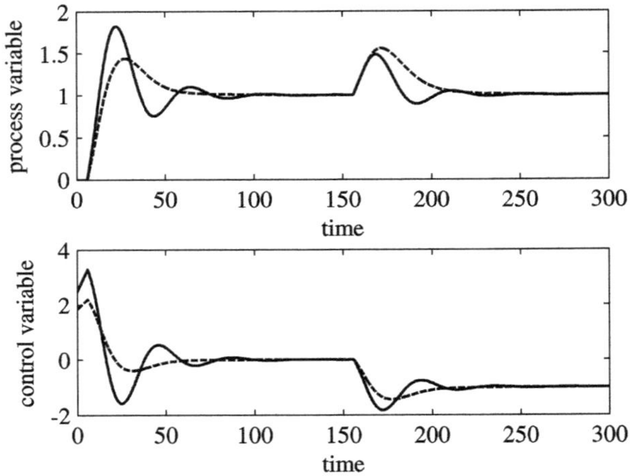

Fig. 2.28 Results obtained with method based on the minimisation of an H_∞ performance index. *Solid line*: $\lambda = 6$. *Dashed line*: $\lambda = 12$

where, according to (2.145), $a = 2\lambda + L$. Thus, by considering $Q(s) = \tilde{Q}(s)F(s)$ and $C(s) = Q(s)/(1 - P(s)Q(s))$ a PI controller is obtained with

$$K_p = \frac{2\lambda + L}{K(\lambda + L)^2}, \tag{2.152}$$

$$T_i = 2\lambda + L. \tag{2.153}$$

By applying the same reasoning, if a SOIPDT model is considered, the filter can be selected as

$$F(s) = \frac{as + 1}{(\lambda s + 1)^3}, \tag{2.154}$$

where $a = 3\lambda + L$ is determined from (2.145). The resulting controller is an output-filtered PID controller (2.4) with

$$K_p = \frac{3\lambda + L + T}{K(3\lambda^2 + 3\lambda L + L^2)}, \tag{2.155}$$

$$T_i = 3\lambda + L + T, \tag{2.156}$$

$$T_d = \frac{(3\lambda + L)T}{3\lambda + L + T}, \tag{2.157}$$

$$T_f = \frac{\lambda^3}{3\lambda^2 + 3\lambda L + L^2}. \tag{2.158}$$

As in IMC, the user-chosen parameter λ can handle the trade-off between aggressiveness and robustness. As an example, if the process (2.58) is considered, the PI controller with $K_p = 2.47$ and $T_i = 18$ is obtained from the tuning rules (2.152)–(2.153) by selecting $\lambda = L = 6$, while for $\lambda = 2L = 12$, the parameters $K_p = 1.83$

and $T_i = 30$ are obtained. Results related to the set-point and load disturbance step response are shown in Figure 2.28.

2.4 Conclusions

In this chapter, the use of PID controllers for the control of integral processes has been addressed. After an introduction on PID controllers, it has been shown that this kind of controllers can be employed effectively for both the set-point following and load disturbance rejection tasks, especially if the control requirements are not too tight. Both open-loop and closed-loop techniques for the estimation of the process parameters have been described. Then, starting from the model obtained, different approaches have been presented for the tuning of the PID parameters with the aim of showing that the tuning problem can be tackled from different viewpoints, each with specific features.

Chapter 3
Stability Region

The determination of the set of the stabilising parameters of a PID controller for an integral process is discussed in this chapter. In particular, the stability region of a PI controller for an IPDT process is analysed first. In this case, in addition to the stability region, the achievable stability margins can also be determined. Then, the case of a stabilising PID controller is considered for both IPDT and SOIPDT processes. The set of parameters is determined by applying the Hermite–Biehler theorem to quasi-polynomials. Note that, instead of presenting all the mathematical details of the procedures employed (for which the reader can refer, for example, to [18, 111, 112], where processes with a first-order-plus-dead-time model are considered), the features related to integral processes are highlighted.

3.1 Stability Region Under the PI Control

If a PI controller is applied to an IPDT process, the set of stabilising parameters can be determined analytically, after a suitable normalisation of the system. The analysis yields also the achievable stability margins (namely, the gain margin and phase margin), and this can be effectively exploited in the tuning of the PI controller [137].

3.1.1 *Normalisation of the System*

Consider the IPDT process (2.14) controlled, in a unity feedback scheme (see Figure 2.1), by the PI controller

$$C(s) = K_p\left(1 + \frac{1}{T_i s}\right),$$

where $K_p > 0$ is the proportional gain, and $T_i > 0$ is the integral time constant.

A. Visioli, Q.-C. Zhong, *Control of Integral Processes with Dead Time*,
Advances in Industrial Control,
DOI 10.1007/978-0-85729-070-0_3, © Springer-Verlag London Limited 2011

The loop transfer function of the system is

$$W(s) = C(s)P(s) = K_p\left(1 + \frac{1}{T_i s}\right)\frac{K}{s}e^{-Ls}, \tag{3.1}$$

which involves four parameters. If the process parameters K and L are normalised into

$$\bar{K}_p = LKK_p \quad \text{and} \quad \bar{T}_i = \frac{T_i}{L}, \tag{3.2}$$

then the loop transfer function is correspondingly normalised as

$$\bar{W}(s) = \bar{K}_p\left(1 + \frac{1}{\bar{T}_i s}\right)\frac{1}{s}e^{-s}. \tag{3.3}$$

This involves only two parameters. As a result, $\bar{W}(s)$ can be regarded as a system with the process

$$\bar{P}(s) = \frac{1}{s}e^{-s}$$

and the controller

$$\bar{C}(s) = \bar{K}_p\left(1 + \frac{1}{\bar{T}_i s}\right),$$

where $\bar{K}_p > 0$ and $\bar{T}_i > 0$ are the normalised proportional gain and the normalised integral time constant, respectively. As a matter of fact, (3.3) can be obtained from (3.1) by substituting s with $\frac{1}{L}s$. This means that the Nyquist plots of $W(s)$ and $\bar{W}(s)$ have the same form but arrive at the same point with different frequencies. Therefore, the design of $C(s)$ for $W(s)$ can be done via designing $\bar{C}(s)$ for $\bar{W}(s)$ and then recovering the parameters of $C(s)$ from (3.2). As can be seen later, this considerably simplifies the system analysis and design.

3.1.2 Stability Region

Rewrite $\bar{W}(s)$ as $\bar{W}(j\omega) = \mathrm{Re}(\omega) + j\mathrm{Im}(\omega)$, where

$$\mathrm{Re}(\omega) = -\frac{\bar{K}_p\cos\omega}{\bar{T}_i\omega^2} - \frac{\bar{K}_p\sin\omega}{\omega},$$

$$\mathrm{Im}(\omega) = \frac{\bar{K}_p\sin\omega}{\bar{T}_i\omega^2} - \frac{\bar{K}_p\cos\omega}{\omega}.$$

As $\omega \to 0$, there are

$$\lim_{\omega\to 0}\mathrm{Re}(\omega) = -\infty, \qquad \lim_{\omega\to 0}\mathrm{Im}(\omega) = \begin{cases} -\infty, & \bar{T}_i > 1, \\ 0, & \bar{T}_i = 1, \\ +\infty, & 0 < \bar{T}_i < 1. \end{cases}$$

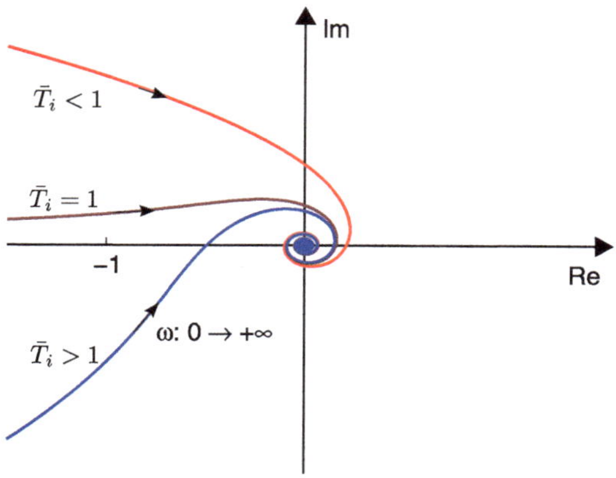

Fig. 3.1 Nyquist plot of the loop transfer function $\bar{L}(s)$ of the system

The Nyquist plots for these cases are shown in Figure 3.1. In the case of $0 < \bar{T}_i < 1$ or $\bar{T}_i = 1$, the Nyquist curve starts above the real axis and encircles the point $(-1, 0)$. In the case of $\bar{T}_i > 1$, it is possible for the Nyquist curve not to encircle the point $(-1, 0)$. Therefore, a necessary condition for the closed-loop system stability is $\bar{T}_i > 1$. This means that the integral time constant of the PI controller $C(s)$ must be greater than the dead time of the process $P(s)$. Furthermore, a necessary and sufficient condition for the stability is given below.

Theorem 3.1 *The closed-loop feedback system with the open-loop transfer function given in* (3.3) *is stable if and only if* $\bar{K}_p$ *and* $\bar{T}_i$ *satisfy*

$$\bar{K}_p < \frac{\bar{T}_i \omega_p^2}{\sqrt{\bar{T}_i^2 \omega_p^2 + 1}}, \tag{3.4}$$

where $\omega_p \in (0, \frac{\pi}{2}]$ *is the solution of*

$$\omega_p = \arctan(\bar{T}_i \omega_p). \tag{3.5}$$

Proof The closed-loop feedback system is stable if and only if the Nyquist curve crosses the real axis from the right side of the point $(-1, 0)$, *i.e.*, satisfying $\text{Re}(\omega) > -1$ when $\text{Im}(\omega) = 0$, which gives

$$\frac{\bar{K}_p \cos \omega_p}{\bar{T}_i \omega_p^2} + \frac{\bar{K}_p \sin \omega_p}{\omega_p} < 1, \tag{3.6}$$

$$\frac{\sin \omega_p}{\bar{T}_i \omega_p} - \cos \omega_p = 0, \tag{3.7}$$

where ω_p is called the phase crossover frequency. Because $\bar{T}_i > 1$, the minimum solution of ω_p in (3.7) lies in the interval $(0, \frac{\pi}{2}]$, where $\frac{\pi}{2}$ is obtained as $\bar{T}_i \to +\infty$ (in this situation, the PI controller degenerates to a P controller). This means that the

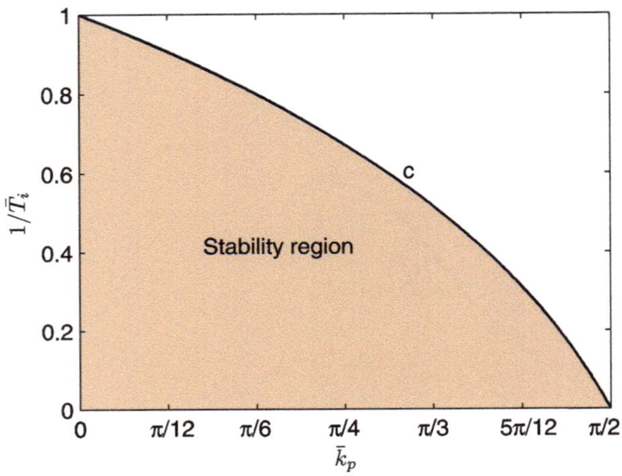

Fig. 3.2 Region of control parameters $\bar{K}_p$ and $\bar{T}_i$ to stabilise the system

Nyquist curve crosses the real axis for the first time at a frequency not more than $\frac{\pi}{2}$. Thus, (3.7) can be converted into (3.5). Furthermore, simplifying (3.6) with (3.7) gives (3.4). □

When the Nyquist curve crosses the point $(-1,0)$, then

$$\bar{K}_p = \frac{\bar{T}_i \omega_p^2}{\sqrt{\bar{T}_i^2 \omega_p^2 + 1}}.$$

From this, together with (3.5), the relationship between $\bar{K}_p$ and $\bar{T}_i$ can be solved numerically, which is shown in Figure 3.2 as the curve c. Note that as $\bar{T}_i \to +\infty$, $\omega_p \to \frac{\pi}{2}$ and $\bar{k}_p \to \frac{\pi}{2}$; as $\bar{k}_p \to 0$, $\omega_p \to 0$ and $\bar{T}_i \to 1$. Obviously, the filled area in Figure 3.2 corresponds to (3.4), which gives the stability region.

3.1.3 Achievable Stability Margins

The well-known gain margin A_m and phase margin ϕ_m are defined as

$$A_m = \frac{1}{|\bar{L}(j\omega_p)|}, \tag{3.8}$$

$$\phi_m = \arg\big[\bar{L}(j\omega_g)\big] + \pi, \tag{3.9}$$

where ω_p is the phase crossover frequency defined in (3.5) and ω_g is the gain crossover frequency defined by

$$\big|\bar{L}(j\omega_g)\big| = 1. \tag{3.10}$$

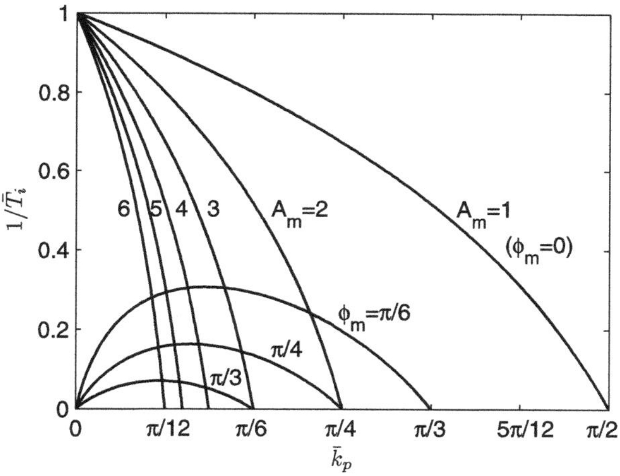

Fig. 3.3 Relationship between stability margins A_m and ϕ_m and control parameters $\bar{K}_p$ and $\bar{T}_i$

Conventionally, A_m, which is larger than 1, is converted so that the gain margin has the unit "dB". Here it is kept dimensionless to simplify the expression and calculation. Before the Nyquist curve reaches the real axis for the first time, $\omega \leq \frac{\pi}{2}$, and $\mathrm{Re}(\omega)$ is always negative, so ϕ_m is in the range of $0 < \phi_m < \frac{\pi}{2}$.

Substituting (3.3) into (3.8) and (3.9) gives

$$A_m = \frac{\bar{T}_i \omega_p^2}{\bar{K}_p \sqrt{\bar{T}_i^2 \omega_p^2 + 1}}, \tag{3.11}$$

$$\phi_m = \arctan(\bar{T}_i \omega_g) - \omega_g, \tag{3.12}$$

where ω_p is given in (3.5), and ω_g, according to (3.10), is

$$\omega_g = \frac{\sqrt{2}}{2} \sqrt{\bar{K}_p^2 + \sqrt{\bar{K}_p^4 + 4\bar{K}_p^2/\bar{T}_i^2}}. \tag{3.13}$$

For a specified gain margin A_m, the parameters $\bar{K}_p$ and $\bar{T}_i$ satisfying (3.11) and (3.5) can be solved numerically. The solutions for typical specifications of $A_m = 2, 3, 4, 5, 6$ are shown in Figure 3.3 and are called gain-margin curves. As $\bar{K}_p \to 0$, $\bar{T}_i \to 1$. This is the point (0, 1). As $\bar{T}_i \to +\infty$, $\bar{K}_p \to \bar{K}_p^A$ with

$$\bar{K}_p^A = \frac{\pi}{2A_m}.$$

This is the intersection point of the gain-margin curve and the horizontal axis. Similarly, for a specified phase margin ϕ_m, the parameters $\bar{K}_p$ and $\bar{T}_i$ satisfying (3.12) and (3.13) can be solved numerically as well. The solutions for typical specifications of $\phi_m = \frac{\pi}{6}, \frac{\pi}{4}, \frac{\pi}{3}$ are also shown in Figure 3.3 and are called phase-margin curves. As $\bar{T}_i \to +\infty$, $\bar{K}_p \to 0$ or $\bar{K}_p \to \bar{K}_p^{\phi}$ with

$$\bar{K}_p^{\phi} = \frac{\pi}{2} - \phi_m.$$

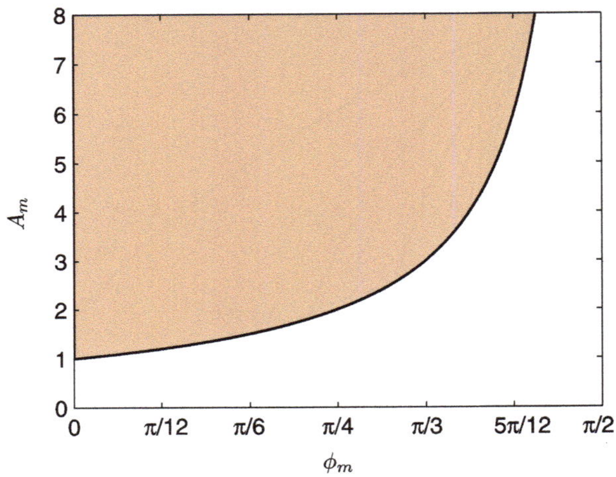

Fig. 3.4 Achievable stability margins A_m and ϕ_m

This is the right-side endpoint of the phase-margin curve. The curve for $A_m = 1$ or $\phi_m = 0$ is the curve c shown in Figure 3.2.

So far, the system for a specified gain margin or phase margin has been designed. When both margins are specified, the parameters of $\bar{C}(s)$ can be obtained from the intersection point of the relevant gain-margin curve and phase-margin curve in Figure 3.3. However, no arbitrary A_m and ϕ_m can be achieved simultaneously. There is a constraint on the achievable stability margins.

Theorem 3.2 *The achievable stability margins A_m and ϕ_m satisfy the following constraint*:

$$\phi_m \leq \frac{\pi}{2}\left(1 - \frac{1}{A_m}\right).$$

Proof For an arbitrary pair (A_m, ϕ_m), there exists either a unique pair $(\bar{K}_p, \bar{T}_i)$ or no pair $(\bar{K}_p, \bar{T}_i)$ to meet them, depending on whether or not the relevant gain-margin and phase-margin curves intersect in Figure 3.3. They intersect with each other when $\bar{K}_p^A \leq \bar{K}_p^\phi$, *i.e.*,

$$\frac{\pi}{2A_m} \leq \frac{\pi}{2} - \phi_m,$$

where the "=" is satisfied as $\bar{T}_i \to +\infty$. This gives the condition in the theorem. □

The achievable stability margins are shown in the filled area of Figure 3.4 (including the curve). As A_m approaches $+\infty$, ϕ_m approaches $\pi/2$.

When the system has been designed for the specified stability margins A_m and ϕ_m, the uncertainties of the process parameters are determined too. Assuming that

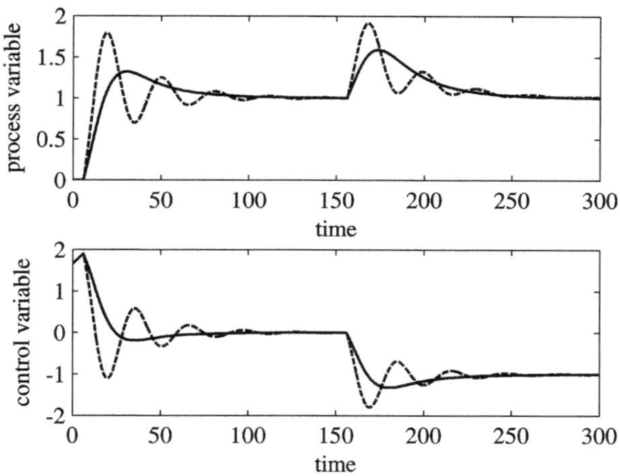

Fig. 3.5 Results obtained with the Ziegler–Nichols tuning rules based on a nonparametric model of the process. *Solid line*: PID controller with no set-point weight. *Dashed line*: PID controller with set-point weight

there exists an uncertainty K_Δ in the process gain K ($K_\Delta + K > 0$), according to (3.2) and the definition of the gain margin, the system is robustly stable if

$$-1 < \frac{K_\Delta}{K} < A_m - 1. \tag{3.14}$$

Similarly, if there exists an uncertainty L_Δ in the process dead time L ($L_\Delta + L \geq 0$), then the system is robustly stable when

$$-1 \leq \frac{L_\Delta}{L} < \frac{\phi_m}{\omega_g}. \tag{3.15}$$

3.1.4 An Illustrative Example

Consider the process (2.58). The gain and phase margins are chosen as $A_m = 3$ and $\phi_m = \frac{\pi}{4}$, as often used in the literature. It can be found that $\bar{K}_p \simeq 0.5$ and $1/\bar{T}_i \simeq 0.15$ from Figure 3.3 and then $K_p = 1.647$ and $T_i = 40$ from (3.2). Simulation results with a unit step input $r(t)$ and a unit step disturbance, acting at $t = 150$ s, are shown in Figure 3.5 as a solid line for the nominal case. From (3.14) and (3.15) it can be determined that the system is robustly stable for $K_\Delta < 0.1$ and $L_\Delta < 0.4$. The response of the system when the process gain has been increased to 0.1 and the dead time has been increased to 6.2 is shown as a dashed line in Figure 3.5, where it appears that the stability is preserved.

3.2 Stability Region Under the PID Control

In case a PID controller is employed, the determination of the set of stabilising parameters can be performed by exploiting a version of the Hermite–Biehler theorem applicable to quasi-polynomials. In particular, the admissible range of the proportional gain is computed first. Then, for each value of the proportional gain in that range, the set of stabilising values of the integral and derivative gain are found. Both the cases of IPDT and SOIPDT processes are addressed hereafter.

3.2.1 IPDT Processes

When an IPDT process is considered, a simple numerical procedure for the determination of the set of stabilising parameter can be employed, as it is illustrated in the following subsection [88].

3.2.1.1 Determination of the Set of Stabilising Parameters

Consider the process (2.14) (where the gain K is assumed to be positive, without loss of generality) controlled by a PID controller which is conveniently written in ideal form

$$C(s) = K_p + \frac{K_i}{s} + K_d s. \tag{3.16}$$

The problem of finding the stabilising set of (K_p, K_i, K_d) values for which the closed-loop system

$$H(s) = \frac{C(s)P(s)}{1 + C(s)P(s)} \tag{3.17}$$

is stable is tackled by considering the characteristic equation of the closed-loop system (3.17):

$$\delta(s) := s^2 + \left(KK_i + KK_p s + KK_d s^2\right)e^{-Ls} = 0, \tag{3.18}$$

which, by multiplying both sides by e^{Ls}, can be rewritten as

$$f(s) := \delta(s)e^{Ls} = s^2 e^{Ls} + KK_i + KK_p s + KK_d s^2 = 0. \tag{3.19}$$

The following theorem [52] will be used hereafter in order to provide a necessary condition for the stabilising set of parameters.

Theorem 3.3 *Consider the quasi-polynomial*

$$\Lambda(s) = \sum_{l=1}^{r} \sum_{i=0}^{n} \lambda_{il} e^{L_i s} s^{n-i}, \tag{3.20}$$

where $L_1 < L_2 < \cdots < L_r$, $L_r + L_1 > 0$, *and the principal term* $\lambda_{0r} \neq 0$. *If* $\Lambda(s)$ *is a stable quasi-polynomial, then the derivative of* $\Lambda(s)$ *with respect to* s *is stable, or, equivalently, if the derivative of* $\Lambda(s)$ *is unstable, then* $\Lambda(s)$ *is unstable.*

Necessary conditions for the stabilising PID controller parameters can now be stated [87].

Theorem 3.4 *Necessary conditions for the PID controller* (3.16) *to stabilise the IPDT process* (2.14) *are*

$$K_p > 0, \qquad K_i > 0, \qquad (1 + KK_d) > 0. \tag{3.21}$$

Proof As e^{Ls} has no finite zeros, the system is stable if and only if the zeros of $f(s)$ are in the open left-half plane. If $s = 0$, Expression (3.19) reduces to $KK_i = 0$, and therefore there exists a singular boundary at $K_i = 0$. The side of the stability boundary $K_i = 0$ that has less unstable poles has therefore to be found. For an arbitrary small value of s, (3.19) can be approximated as

$$KK_p s + KK_i \simeq 0, \tag{3.22}$$

i.e., $s \simeq -K_i/K_p$. This implies that for $K_p > 0$, the region for $K_i > 0$ has one less unstable pole than for $K_i > 0$ (note that, by applying a similar reasoning, $K_p < 0$ implies $K_i < 0$). Then, the first derivative of (3.19) can be computed as

$$f'(s) = (2s + Ls^2)e^{Ls} + KK_p + 2KK_d s = 0. \tag{3.23}$$

For $s = 0$, Expression (3.23) yields $K_p = 0$, and by applying the same reasoning as before, it can be deduced that if $KK_p/(1 + KK_d) < 0$, then $f'(s)$ is unstable, and therefore also $f(s)$ is unstable by Theorem 3.3. The second derivative of (3.19) can be computed as

$$f''(s) = (2 + 4Ls + L^2s^2)e^{Ls} + 2KK_d = 0. \tag{3.24}$$

For $s = 0$, Expression (3.24) yields $1 + KK_d = 0$. If it is assumed that $K_d = -1/K + \varepsilon$, where ε is an arbitrarily small positive number, then (3.24) is approximated as

$$4Ls + 2 + 2K\left(-\frac{1}{K} + \varepsilon\right) = 0, \tag{3.25}$$

i.e., $s \simeq -K\varepsilon/(2L)$ (which is less than zero). This implies that the quasi-polynomial $f''(s)$ has one less right-half plane zero in the region $K_d > -1/K$, that is, by Theorem 3.3, a necessary condition for the system to be stable is $1 + KK_d > 0$. By taking into account all the considerations done, the inequalities (3.21) are trivially demonstrated. □

Then, the Hermite–Biehler theorem can be employed in order to determine the stabilising set of (K_p, K_i, K_d). By substituting $s = j\omega$ into (3.19) it results

$$f(\omega) = -\omega^2 e^{jL\omega} + KK_i + jKK_p\omega - \omega^2 KK_d = 0. \tag{3.26}$$

By denoting as $f_r(\omega)$ and $f_i(\omega)$ the real and imaginary parts of $f(\omega)$, respectively, and by $f_r'(\omega)$ and $f_i'(\omega)$ their first-order derivatives with respect to ω, the following well-known theorem can be stated [10, 52, 100].

Theorem 3.5 *The quasi-polynomial in* (3.26) *is stable if and only if*

1. $f_r(\omega)$ *and* $f_i(\omega)$ *have only real roots, and these roots interlace;*
2. $f_i'(\omega^*)f_r(\omega^*) - f_i(\omega^*)f_r'(\omega^*) > 0$ *for some* $\omega^* \in (-\infty, \infty)$ *(increasing phase condition).*

The following theorem is employed to verify that $f_r(\omega)$ and $f_i(\omega)$ have only real roots [10].

Theorem 3.6 *Let* p *and* q *denote the highest powers of* s *and* $e^{\theta s}$*, respectively, in* $f(s)$*. Let* η *be a constant such that the coefficients of terms of highest degree in* $f_r(\omega)$ *and* $f_i(\omega)$ *do not vanish at* $\omega = \eta$*. Then, a necessary and sufficient condition under which* $f_r(\omega)$ *or* $f_i(\omega)$ *have only real roots is that in the strip* $-2l\pi + \eta \le \omega \le 2l\pi + \eta$*,* $f_r(\omega)$ *or* $f_i(\omega)$ *has exactly* $4lq + p$ *real roots for a sufficiently large* l*.*

Theorem 3.6 can be employed to determine the admissible range of K_p for which $f_i(z)$ has only real roots. By taking $l = 1$, as $p = 2$ and $q = 1$, there have to be six real roots in the strip $-2\pi + \eta \le \omega \le 2\pi + \eta$. By taking $z = L\omega$, it is

$$f_i(z) = KK_p\frac{z}{L} - \frac{z^2}{L^2}\sin(z) = 0, \tag{3.27}$$

for which it is apparent that a solution $z_0 = 0$ occurs. Then, it can be trivially determined that

$$\bar{f}_i(z) := KK_p - \frac{z}{L}\sin(z) = 0, \tag{3.28}$$

which is an even function. A plot of an example of function $\bar{f}_i(z)$ is shown in Figure 3.6, where it emerges that the roots z_1 and z_2 locate in the interval $[0, \pi]$, and, in general, the roots z_k and z_{k+1}, $k = 3, 5, \ldots$, are located in the interval $[(k-1)\pi, k\pi]$. In addition, the root z_k gets closer to $(k-1)\pi$ as k increases. Thus, by denoting d as the distance between 2π and z_3, it is sufficient to take $\eta = d + \varepsilon$ with an arbitrary small ε in order to have six roots (namely, $z_{-2}, \ldots, z_3$) in the strip $[-2l\pi + \eta, 2l\pi + \eta]$. In general, it is sufficient to take $d = z_{2l+1} - 2\pi l$ in order to have, for any l, $4l + 2$ roots in the strip $[-2l\pi + \eta, 2l\pi + \eta]$ where $\eta = d + \varepsilon$.

Note that the coefficient of the term of highest degree in $f_i(z)$ (namely, $\sin(z)$), does not vanish at $z = \eta$, because, as mentioned above, the root z_k gets closer to $(k-1)\pi$ as k increases, and therefore, for an arbitrarily small ε, it is guaranteed that z_1 is greater than $d + \varepsilon$.

The admissible range of K_p, namely, the admissible maximum value K_{pm}, can be found by determining the condition for which z_1 and z_2 are coincident, because this is the limit case for which the hypotheses of Theorem 3.6 hold. It is necessary to find K_{pm} such that $\bar{f}_i(z)$ is tangent to the abscissa axis, namely, $\bar{f}_i(z) = 0$ and $\bar{f}_i'(z) = 0$. Thus, the following system results:

$$\begin{cases} \dfrac{KK_{pm}}{\sin(z)} = \dfrac{z}{L}, \\ -\dfrac{KK_{pm}\cos(z)}{\sin^2(z)} = \dfrac{1}{z}, \end{cases} \tag{3.29}$$

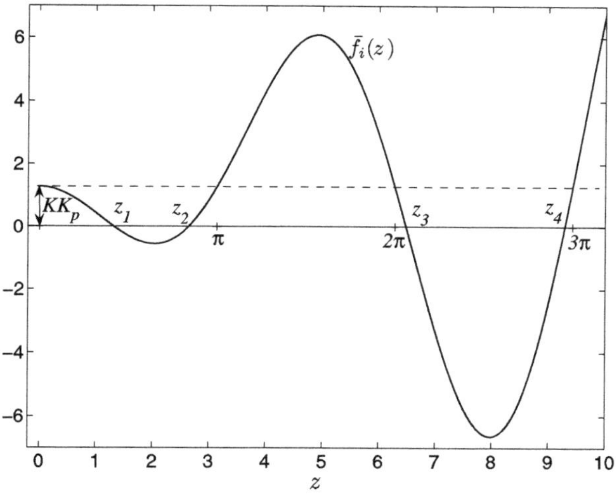

Fig. 3.6 Example of function $\bar{f}_i(z)$

which yields

$$K_{pm} = \frac{1}{K}\frac{\alpha}{L}\sin(\alpha), \tag{3.30}$$

where α is the solution of the equation

$$z = -\tan(z). \tag{3.31}$$

Now the interlacing of the roots of $f_i(z)$ and $f_r(z)$ is addressed (see condition 1 of Theorem 3.5). Consider

$$f_r(z) = KK_i - KK_d\frac{z^2}{L^2} - \frac{z^2}{L^2}\cos(z). \tag{3.32}$$

Then, $f_r(z_0) > 0$ because $z_0 = 0$ and $f_r(0) > 0$ because $K_i > 0$ (see Theorem 3.4). Thus, it has to be

$$f_r(z_1) < 0, \quad f_r(z_2) > 0, \quad \ldots, \tag{3.33}$$

and $f_r(z)$ can be rewritten as

$$f_r(z) = \frac{Kz^2}{L^2}\big[-K_d + a(z)K_i + b(z)\big], \tag{3.34}$$

where

$$a(z) = \frac{L^2}{z^2}, \qquad b(z) = -\frac{1}{K}\cos(z). \tag{3.35}$$

Thus, condition (3.33) can be rewritten as

$$(-1)^k K_d < (-1)^k a(z_k)K_i + (-1)^k b(z_k), \quad k = 1, 2, \ldots. \tag{3.36}$$

It can be noted now that in the range $(k\pi, (k+2)\pi)$, $f_r(z)$ has one maximum and one minimum, and in the same range, there are the two roots z_{2k+1} and z_{2k+2} of

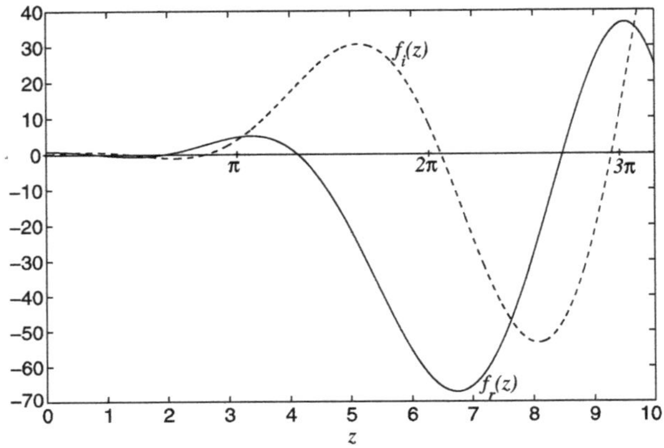

Fig. 3.7 Example of functions $f_i(z)$ and $f_r(z)$

$f_i(z)$. If condition (3.36) is satisfied (namely, $f_r(z_{2k+1}) < 0$ and $f_r(z_{2k+2}) > 0$), there is at least one root of $f_r(z)$ between z_{2k+1} and z_{2k+2}. This root is unique because $f_r(z)$ changes its sign just once between an extremum and the next one. In fact, the number of extrema and of roots in the range $[k\pi, (k+2)\pi]$ is the same (see Figure 3.7).

It is worth noting at this point that the interlacing property and the fact that the roots of $f_i(z)$ are real implies that the roots of $f_r(z)$ are also real [111].

Condition 2 of Theorem 3.5 can be easily proved. In fact, from (3.27) and (3.32) we have

$$f_i'(z) = \frac{KK_p}{L} - \frac{1}{L^2}\big[\cos(z)z^2 + 2\sin(z)z\big], \tag{3.37}$$

and it is therefore sufficient to select $z = 0$ to have

$$f_i'(0)f_r(0) - f_i(0)f_r'(0) = \frac{KK_p}{L}KK_i > 0. \tag{3.38}$$

Summarising, for each value of K_p in $(0, K_{pm})$, the intersection of the set of inequalities (3.36) determines the set of (K_i, K_d) values for which the roots of $f_i(z)$ and $f_r(z)$ interlace and therefore that stabilise the closed-loop system (3.17). Note that (3.36) represents, for each value of z_k, a half-plane in the Cartesian coordinate system (K_i, K_d). Indeed, if the set of inequalities (3.36) is substituted with the set of equalities

$$(-1)^k K_d = (-1)^k a(z_k)K_i + (-1)^k b(z_k), \quad k = 1, 2, \dots, \tag{3.39}$$

then, (3.39) represents, for each value of z_k, a straight line in the Cartesian coordinate system (K_i, K_d).

Thus, for a fixed value of K_p in $(0, K_{pm})$, each (odd) root z_{2k+1} of $f_i(z)$ with $k \in \mathbb{N}$ determines a straight line with slope $a(z_{2k+1})$ and y-intercept $b(z_{2k+1})$. As k increases, the slope decreases to zero, and the y-intercept tends to $-1/K$ because $(z_{2k+1} - 2k\pi)$ is a positive succession that converges monotonically to zero and

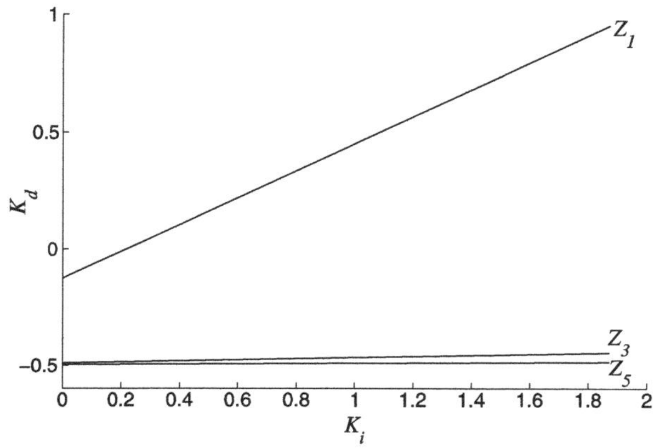

Fig. 3.8 Straight lines determined by the odd roots z_{2k+1}

therefore $\cos(z_{2k+1})$ converges to one. Indeed, $b(z_{2k+1})$ is strictly monotonically decreasing as k increases because $\cos(z)$ is strictly monotonically decreasing in the range $(2k\pi, (2k+1)\pi)$ and therefore

$$\cos(z_1) < \cos(z_3) < \cos(z_5) < \cdots, \tag{3.40}$$

which implies

$$-\frac{1}{K}\cos(z_1) > -\frac{1}{K}\cos(z_3) > -\frac{1}{K}\cos(z_5) > \cdots. \tag{3.41}$$

Thus, with respect to the (odd) roots z_{2k+1}, the root z_1 can only be taken into account. The situation is exemplified in Figure 3.8.

Regarding the (even) roots z_{2k} of $f_i(z)$ with $k \in \mathbb{N}$, denote as $(\tilde{x}_k, \tilde{y}_k)$ the intersection of the two straight lines (3.39) obtained for z_{2k} and z_{2k+2}. Then, the following theorem can be stated.

Theorem 3.7 *As $k \to \infty$, $(\tilde{x}_k, \tilde{y}_k)$ converges to the point $(\frac{KK_p^2}{2}, \frac{1}{K})$ by monotonically decreasing.*

Proof Because $(z_{2k} - (2k-1)\pi)$ is a negative succession that converges monotonically to zero as k increases, it can be determined that

$$\cos(z_2) > \cos(z_4) > \cos(z_6) > \cdots, \tag{3.42}$$

which implies

$$-\frac{1}{K}\cos(z_2) < -\frac{1}{K}\cos(z_4) < -\frac{1}{K}\cos(z_6) < \cdots. \tag{3.43}$$

Because

$$\lim_{k\to+\infty}\big(z_{2k} - (2k-1)\pi\big) = 0, \tag{3.44}$$

it can be deduced trivially that

$$\lim_{k \to +\infty} \cos(z_{2k}) = -1 \tag{3.45}$$

and therefore

$$\lim_{k \to +\infty} b(z_{2k}) = \frac{1}{K}. \tag{3.46}$$

By also taking into account that

$$\lim_{k \to +\infty} a(z_{2k}) = 0, \tag{3.47}$$

it can be determined that for $k \to +\infty$, a horizontal straight line $K_d = 1/K$ results. Hence,

$$\lim_{k \to +\infty} \tilde{y}_k = \frac{1}{K}. \tag{3.48}$$

Starting from (3.44), the following expression can also be written:

$$\bar{f}_i(z_{2k}) = K K_p - \frac{1}{\theta} z_{2k} \big[\varepsilon + o(\varepsilon)\big], \tag{3.49}$$

where

$$\varepsilon := (2k-1)\pi - z_{2k} \tag{3.50}$$

because $\varepsilon \to 0$ as $k \to +\infty$ and therefore $\sin(\varepsilon) \sim \varepsilon$. Hence, for $k \to +\infty$, $\varepsilon \to 0$ as $(K K_p \theta)/z_{2k}$. When $(z_{2k} - (2k-1)\pi) \to 0$, $\cos(z_{2k})$ can be expressed in Taylor series as

$$\cos(z_{2k}) = -1 + \frac{\varepsilon^2}{2} + o\big(\varepsilon^2\big), \tag{3.51}$$

which, by taking into account (3.50), yields

$$\cos(z_{2k}) = -1 + \frac{K^2 K_p^2 \theta^2}{2 z_{2k}^2} + o\left(\frac{1}{z_{2k}^2}\right), \tag{3.52}$$

and therefore $b(z_{2k})$ increases as $\frac{1}{K}(1 - \frac{K^2 K_p^2 \theta^2}{2 z_{2k}^2})$ as $k \to +\infty$. By considering that $\tilde{x}$ is given by

$$\tilde{x}_k = \frac{b(z_{2k}) - b(z_{2k+2})}{a(z_{2k+2}) - a(z_{2k})}, \tag{3.53}$$

by taking into account (3.52) it can be written that

$$\lim_{k \to +\infty} \tilde{x} = \lim_{k \to +\infty} \frac{\frac{1}{K}\left(1 - \frac{K^2 K_p^2 \theta^2}{2 z_{2k}^2}\right) - \frac{1}{K}\left(1 - \frac{K^2 K_p^2 \theta^2}{2 z_{2k+2}^2}\right)}{\frac{\theta^2}{z_{2k+2}^2} - \frac{\theta^2}{z_{2k}^2}} = \frac{K K_p^2}{2}. \tag{3.54}$$

Thus, from (3.48) and (3.54) it can be deduced that, as $k \to +\infty$, the intersection of the straight lines (3.36) with z_{2k} and z_{2k+2}, $k \in \mathbb{N}$, converges to the point $(\frac{K K_p^2}{2}, \frac{1}{K})$.

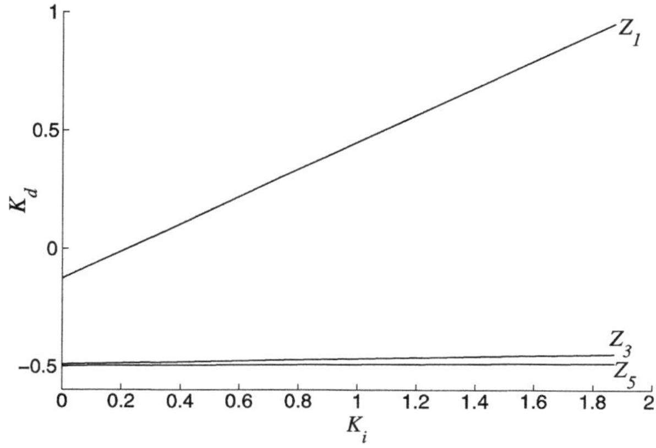

Fig. 3.9 Straight lines determined by the even roots z_{2k}

Both the values of the abscissa and ordinate of the intersection between two successive straight lines decrease (see Figure 3.9). In fact, the value of $\frac{d}{dz}(-\sin(z_{2k}))$ increases as k increases. Thus, by taking into account that $b(z_{2k})$ increases as $\frac{1}{K}(1-\frac{K^2K_p^2\theta^2}{2z_{2k}^2})$ as $k \to +\infty$, it can be deduced that, as k increases, the term $b(z_{2k})$ increases faster than $1-1/z_{2k}^2$, namely, faster than the velocity of variation of the slope $a(z_{2k}) = \theta^2/z_{2k}^2$. □

From the above analysis it can be concluded that, for each K_p in $(0, K_{pm})$, the set of (K_i, K_d) values that stabilise the closed-loop system (3.17) can be determined by considering the intersections of the straight lines for z_1, z_2 and the straight line $K_d = -1/K$ (which corresponds to the straight line for z_{2k} for $k \to +\infty$).

3.2.1.2 An Illustrative Example

In order to illustrate the results presented in Section 3.2.1.1, the same IPDT process (2.58) has been considered. The stabilising regions of (K_i, K_d) for the different stabilising values of $K_p \in (0, K_{pm})$ are shown in Figure 3.10. The associated limits for the controller parameters (*i.e.*, their minimum and maximum limits) are reported in Table 3.1. The stabilising region of (K_i, K_d) for the particular case $K_p = K_{pm}/2 = 2.9969$ is plotted in Figure 3.14. It can be noted that the upper edge of the stability area is almost everywhere horizontal. However, in the left side of this edge, the intersection between the straight line associated with the first zero and the straight line $K_d = 1/K$ can be seen.

Finally, the set-point and load disturbance step response corresponding to the controller parameters $K_p = 2.9969$, $K_i = 0.2582$, and $K_d = 9.6930$ which are in the stability region is shown in Figure 3.15.

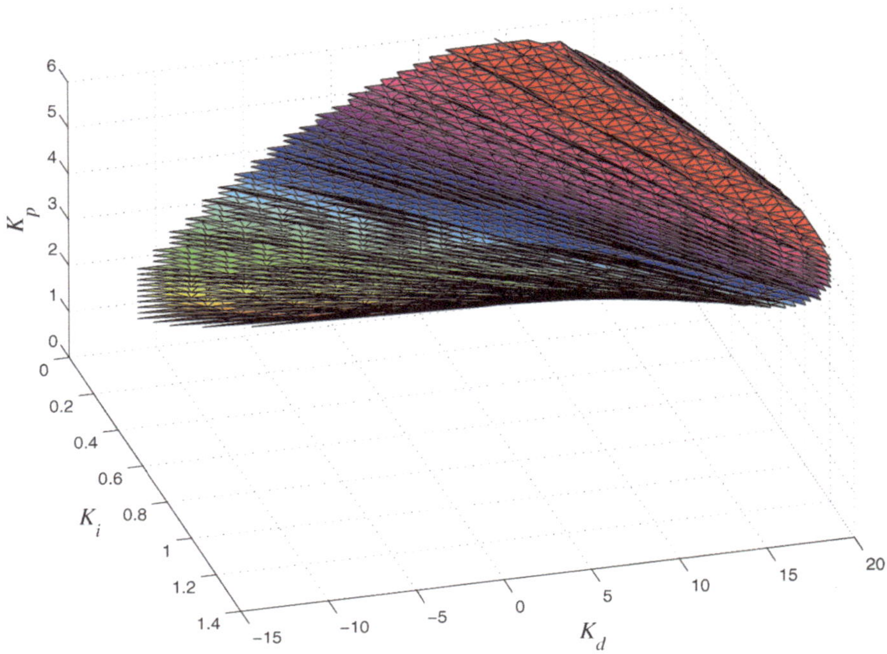

Fig. 3.10 The stabilising regions of (K_i, K_d) for $0 < K_p < K_{pm}$

Table 3.1 The minimum and maximum limits of the controller parameters

Parameter	Minimum value	Maximum value
K_p	0	5.9938
K_i	0	2.0558
K_d	−19.763	26.719

3.2.2 SOIPDT Processes

A more complex procedure has to be employed if an SOIPDT process is considered, as it is illustrated hereafter.

3.2.2.1 Determination of the Set of Stabilising Parameters

A procedure for the determination of the stabilising set of PID parameters has been also proposed for SOIPDT processes with transfer function [87]

$$P(s) = \frac{K}{s(Ts+1)}e^{-Ls}. \tag{3.55}$$

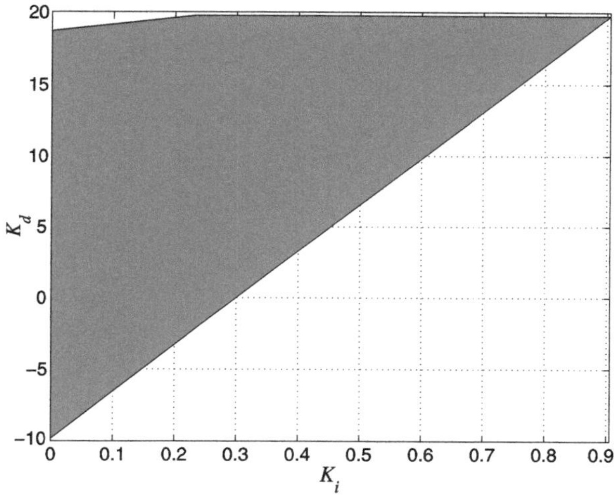

Fig. 3.11 The stabilising region of (K_i, K_d) for $K_p = K_{pm}/2 = 0.31842$

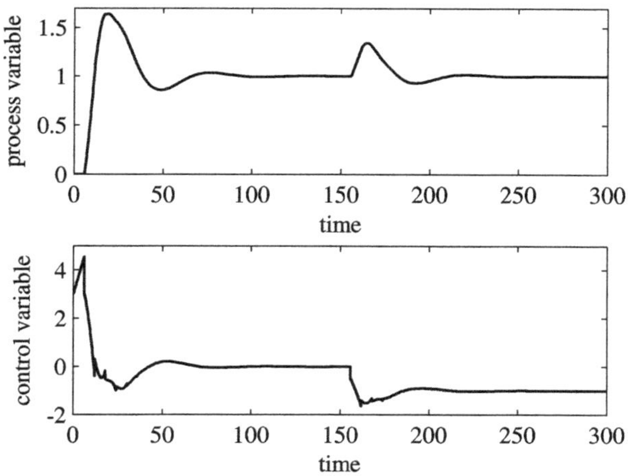

Fig. 3.12 The set-point and load disturbance step response for $K_p = 2.9969$, $K_i = 0.2582$, and $K_d = 9.6930$

In this case the characteristic equation can be written as

$$f(s) := \delta(s)e^{Ls} = s^2(Ts+1)e^{Ls} + KK_i + KK_ps + KK_ds^2 = 0; \quad (3.56)$$

its first derivative is

$$f'(s) = \left(2s + 3Ls^2 + Ls^2 + TLs^3\right)e^{Ls} + KK_p + 2KK_ds, \quad (3.57)$$

and its second derivative is

$$f''(s) = \left[2 + (4L+6T)s + \left(L^2 + 6LT\right)s^2 + TL^2s^3\right]e^{Ls} + 2KK_d. \quad (3.58)$$

By applying the same considerations done in Section 3.2.1.1, Theorem 3.4 holds also for SOIPDT processes.

Then, by taking into account that

$$f_i(z) = KK_p\frac{z}{L} - \frac{z^2}{L^3}\big[Tz\cos(z) + L\sin(z)\big] = 0 \tag{3.59}$$

and

$$f_r(z) = KK_i - KK_d\frac{z^2}{L^2} - \frac{z^2}{L^2}\cos(z) + Tz^3\frac{\sin(z)}{L^3}, \tag{3.60}$$

Theorem 3.6 can be proven by selecting $\eta = 0$ and by considering that, in addition to $z_0 = 0$, the roots of $f_i(z)$ are the solutions of the equation

$$\bar{f_i}(z) := KK_p - T\cos(z)\frac{z^2}{L^2} - \frac{z}{L}\sin(z) = 0, \tag{3.61}$$

which can be rewritten as

$$\frac{KK_p - T\frac{z}{L^2}^2\cos(z)}{\sin(z)} = \frac{z}{L}. \tag{3.62}$$

By noting that $\bar{f_i}(z)$ is an even function, the condition that $f_i(z)$ has only real roots is equivalent to the condition that $\bar{f_i}(z)$ has $2l+1$ real roots in the strip $0 < z \leq 2\pi l$. By analysing the left-hand side of Equation (3.62), denoted as $\Gamma(z)$, it can be derived (see [86] for details) that this condition is met only if the plot of $\Gamma(z)$ intersects the line z/L twice in the interval $(0, \pi)$. This occurs if

$$\frac{KK_pL^2 - T\alpha^2\cos(\alpha)}{L\sin(\alpha)} < \alpha, \tag{3.63}$$

where α is the solution of the equation

$$\frac{d}{dz}\Gamma(z) = \frac{1}{L}, \tag{3.64}$$

that is,

$$\tan(\alpha) = \frac{2T\alpha + \alpha L}{T\alpha^2 - L}. \tag{3.65}$$

The admissible range of K_p, namely, the admissible maximum value K_{pm}, is therefore determined when the two intersections are coincident, namely, for each value of K_p in the interval $(0, K_{pm}]$, Theorem 3.5 has to be addressed. By following the same approach of Section 3.2.1.1, the real part $f_r(z)$ (see (3.60)), which is an even function, can be written as

$$f_r(z) = \frac{Kz^2}{L^2}\big[-K_d + a(z)K_i + b(z)\big], \tag{3.66}$$

where

$$a(z) = \frac{L^2}{z^2} \tag{3.67}$$

and

$$b(z) = -\frac{1}{K}\cos(z) + \frac{Tz}{KL}\sin(z). \tag{3.68}$$

For $z_0 = 0$, $f_r(z_0) = KK_i > 0$ because $K_i > 0$. Thus, the interlacing condition is

$$f_r(z_1) < 0, \quad f_r(z_2) > 0, \quad \ldots, \tag{3.69}$$

which can be rewritten as

$$(-1)^j K_d < (-1)^j a(z_j)K_i + (-1)^j b(z_j), \tag{3.70}$$

where $j = 1, 2, 3, \ldots$, and z_j's are the positive real roots of (3.59) arranged in increasing order of magnitude. From (3.67) it appears clearly that $a(z_j) > 0$ for every j and $a(z_j) > a(z_{j+1})$. Further, $a(z_j) \to 0$ as $j \to +\infty$. Then, by analysing the function $\Gamma(z)$, it can be noted that if $K_{pm} \le 4\pi^2 T/(KL^2)$, the root z_1 is in the interval $(0, \pi/2)$, and the other (odd) roots z_{2j+1} are in the interval $((2j-0.5)\pi, 2j\pi)$, and they get closer to $(2j-0.5)\pi$ as j increases. Thus, it can be deduced that $1 > \cos(z_1) > \cos(z_3) > \cdots$ and $\lim_{j\to+\infty} \cos(z_{2j+1}) = 0$. Similarly, the (even) roots z_{2j} are in the interval $((2j-1.5)\pi, (2j-1)\pi)$, and they get closer to $(2j-1.5)\pi$ as j increases. Thus, it can be deduced that $-1 < \cos(z_2) < \cos(z_4) < \cdots$ and $\lim_{j\to+\infty}\cos(z_{2j}) = 0$. For the odd roots, $Tz_{2j+1}/(Kl)\sin(z_{2j+1}) \to -\infty$ as $j \to +\infty$ and for the even roots, $Tz_{2j}/(Kl)\sin(z_{2j}) \to +\infty$ as $j \to +\infty$. Further, in both cases, the difference between two successive values of the cosine terms are much smaller than the difference between two successive values of the term $Tz/(Kl)\sin(z)$. Thus, $b(z_{2j+1}) > b(z_{2j+3})$ and $b(z_{2j+1}) \to -\infty$ as $j \to +\infty$, and $b(z_{2j}) < b(z_{2j+2})$ and $b(z_{2j}) \to +\infty$ as $j \to +\infty$. By denoting as $K_i(z_1, z_{2j})$ the value of the abscissa of the intersection point of the straight lines $K_d(z_1)$ and $K_d(z_{2j})$, it can be determined that

$$K_i(z_1, z_{2j}) = \frac{b(z_{2j}) - b(z_1)}{a(z_1) - a(z_{2j})} \tag{3.71}$$

increases as j increases, because the ascending magnitude of the numerator is much larger than the ascending magnitude of the denominator (note that they are both positive). Thus, for $K_{pm} \le 4\pi^2 T/(KL^2)$, the set of inequalities (3.70) can be simplified to

$$K_i > 0, \qquad K_d > a(z_1)K_i + b(z_1), \qquad K_d < a(z_2)K_i + b(z_2). \tag{3.72}$$

By following a similar reasoning, for the case $4\pi^2 T/(KL^2) \le K_{pm} \le (2q\pi)^2 T/(KL^2)$ where $q = 2, 3, 4, \ldots$, it can be determined again that $b(z_{2j+1}) > b(z_{2j+3})$ and $b(z_{2j+1}) \to -\infty$ as $j \to +\infty$ and $b(z_{2j}) < b(z_{2j+2})$ and $b(z_{2j}) \to +\infty$ as $j \to +\infty$. Further, it can be determined that $K_i(z_1, z_{2j}) < K_i(z_1, z_{2j+2})$ for values of j not less than $\min(q)$. In this case, the set of inequalities (3.70) can be rewritten as

$$K_i > 0, \qquad K_d > a(z_1)K_i + b(z_1), \qquad K_d < a(z_{2j})K_i + b(z_{2j}), \tag{3.73}$$

where $j = 1, \ldots, \min(q)$.

Table 3.2 The minimum and maximum limits of the controller parameters

Parameter	Minimum value	Maximum value
K_p	0	4.3434
K_i	0	14.0545
K_d	−1	4.3058

Condition 2 of Theorem 3.5 can be easily proved by selecting $z = 0$ for which, as in (3.38),

$$f_i'(0)f_r(0) - f_i(0)f_r'(0) = \frac{KK_p}{L}KK_i > 0. \tag{3.74}$$

Summarising these results, the following theorem can be stated [87].

Theorem 3.8 *For a given K_p in $(0, K_{pm})$, if $K_{pm} \leq 4\pi^2T/(KL^2)$, the stabilising region in the $K_i - K_d$ plane is a triangle defined by the straight lines $K_i > 0$, $K_d > a(z_1)K_i + b(z_1)$, and $K_d < a(z_2)K_i + b(z_2)$. If $K_{pm} \in (4\pi^2T/(KL^2), (2q\pi)^2T/(KL^2))$, then the stabilising region in the $K_i - K_d$ plane is the area defined by the straight lines $K_i > 0$, $K_d > a(z_1)K_i + b(z_1)$, and $K_d < a(z_{2j})K_i + b(z_{2j})$, where $j = 1, \ldots, \min(q)$.*

It is worth noting that, for the case of SOIPDT processes, the numerical procedure for the determination of the stabilising region is more complex than for IPDT processes.

3.2.2.2 An Illustrative Example

As a illustrative example, consider the process

$$P(s) = \frac{1}{s(s+1)}e^{-0.5s}. \tag{3.75}$$

After having determined the maximum limit of the proportional gain $K_{pm} = 4.3434$, the stabilising regions of (K_i, K_d) for the different stabilising values of $K_p \in (0, K_{pm})$ have been determined. They are shown in Figure 3.13. The associated limits for the controller parameters (*i.e.*, their minimum and maximum limits) are reported in Table 3.1. The stabilising (triangular) region of (K_i, K_d) for the particular case $K_p = K_{pm}/2 = 2.1717$ is plotted in Figure 3.11.

Finally, the set-point and load disturbance step response corresponding to the controller parameters $K_p = 2.1717$, $K_i = 1.9113$, and $K_d = 2.4620$ which are in the stability region is shown in Figure 3.12.

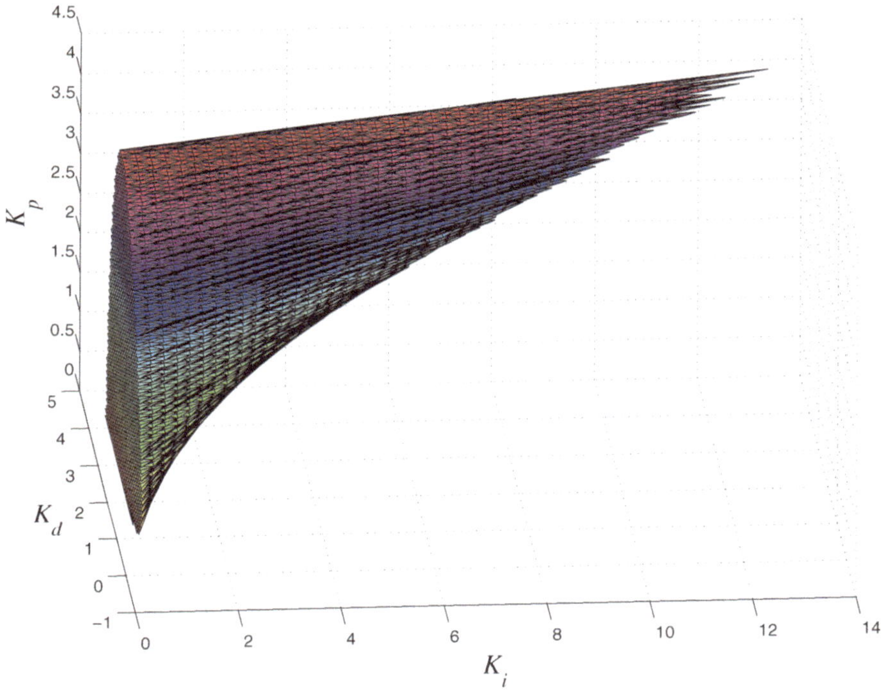

Fig. 3.13 The stabilising regions of (K_i, K_d) for $0 < K_p < K_{pm}$

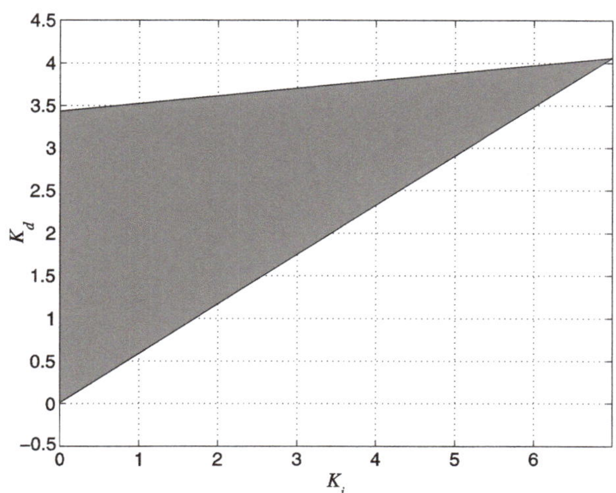

Fig. 3.14 The stabilising region of (K_i, K_d) for $K_p = K_{pm}/2 = 2.1717$

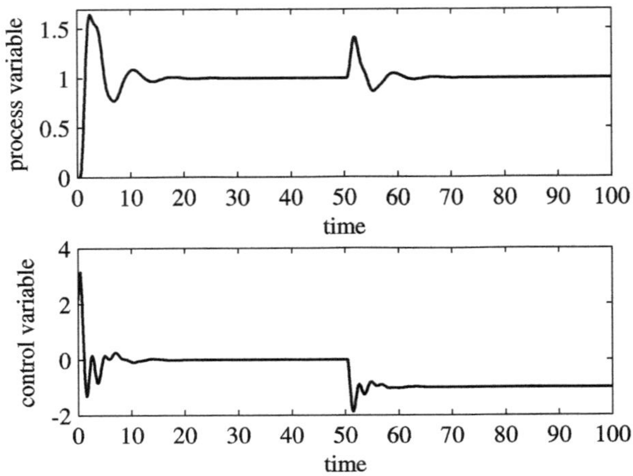

Fig. 3.15 The set-point and load disturbance step response for $K_p = 2.1717$, $K_i = 1.9113$, and $K_d = 2.4620$

3.3 Conclusions

In this chapter, the determination of the stabilising region of PI and PID controllers for IPDT and SOIPDT processes has been addressed. It has been shown that different analyses can be performed depending on the considered process and the considered controller and that in any case the solution can be obtained by simple and computationally efficient numerical procedures. The obtained stabilising regions can be exploited for the development of techniques for the analysis and design of PID controllers for integral processes.

Chapter 4
Performance Assessment and Controller Retuning

In this chapter a methodology for the performance assessment and retuning of PID controllers applied to integral processes is presented. In particular, the deterministic performance related to both the set-point following and load disturbance rejection task is addressed. Routine operating data are employed in order to evaluate the response obtained with given controller parameters with respect to that achievable with appropriate tuning rules. If the performance is not satisfactory, then the controller is retuned suitably.

4.1 Introduction

As shown in Chapter 2, there are many PID controller tuning rules that have been devised for integral processes. However, it is also recognised that in many practical cases PID controllers are poorly tuned because of the lack of time and of the lack of skill of the operator. Actually, as in large plants there are hundreds of control loops, it is almost impossible for operators to monitor each of them manually. For these reasons, it is important to have automatic tools that are first able to assess the performance of a control system and, in case it is not satisfactory, to suggest the way to solve the problem (for example, if a bad controller tuning is detected, then new appropriate values of controller parameters are determined). In this context, it is much appreciated that the retuning is accomplished by using routine operating data (the same that have been employed for the purpose of performance assessment), without the need of performing special experiments (as in the case of standard automatic tuning methodologies) that would lead to time and energy consumption and, in general, would affect the process operations.

Many performance assessment methodologies have been proposed in the literature and successfully applied in industrial settings [45]. In general, although the proposed techniques can be viewed under the same framework (see [38] and references therein), they are generally divided into two categories [104]: *stochastic performance monitoring* in which the capability of the control system to cope with stochastic disturbances is of main concern (works that fall into this class mainly rely

A. Visioli, Q.-C. Zhong, *Control of Integral Processes with Dead Time*, Advances in Industrial Control,
DOI 10.1007/978-0-85729-070-0_4, © Springer-Verlag London Limited 2011

on the concept of minimum variance control [34]), and *deterministic performance monitoring* in which performances related to more traditional design specifications, such as set-point and load rejection disturbance step response parameters, are taken into account [25]. When deterministic requirements are considered, it is realised that an unsatisfactory performance can be caused by different factors [91]. Thus, there is the need to integrate different techniques, each of them devoted to deal with a particular situation.

Restricting the analysis to the tuning assessment of PID controllers, an iterative solution method for the determination of the minimum variance PID controller has been proposed in [53], where there are no assumptions on the modelling of the process. Regarding deterministic performance monitoring, the achievable optimal performance in terms of integrated absolute error for the set-point response has been investigated in [39]. Its knowledge can be exploited to evaluate the performance of an employed PI or PID controller. In [118] the set-point following performance of a PI controller is assessed by taking into account the step response that can be obtained by selecting the controller parameters by means of the Internal Model Control tuning rule. A procedure, which requires a special experiment, for the retuning of a PI controller has been presented in [119].

Regarding load disturbance rejection performance, a methodology to detect sluggish control loops has been presented in [31] and further discussed in [55]. This has also been exploited in [131], where the technique presented assesses the tuning of a PI controller and then gives guidelines on how to retune it, if necessary. A comprehensive description of all these techniques is available in [132].

In the following sections, a methodology for integral processes, based on routine operation data, which assesses the performance of a PID controller and provides a new tuning of all the controller parameters is described. It is worth noting that a similar procedure has also been devised for self-regulating processes [126].

4.2 Problem Formulation

The unity-feedback control system of Figure 4.1 is considered, where the integral process P is controlled by a PID controller whose transfer function is in series (interacting) form (2.5), which is reported hereafter for convenience:

$$C(s) = K_p\left(\frac{T_i s + 1}{T_i s}\right)(T_d s + 1). \tag{4.1}$$

The series form is chosen for the sake of simplicity, as the tuning rules that will be employed in the methodology apply directly to this form. However the use of other forms is straightforward by suitably applying conversion formulae to determine the values of the parameters [132]. Note also that the use of a first-order filter that makes the controller transfer function proper has been neglected for the sake of clarity, but it can be easily selected so that it does not influence the PID controller dynamics significantly (and it filters the high-frequency noise at the same time).

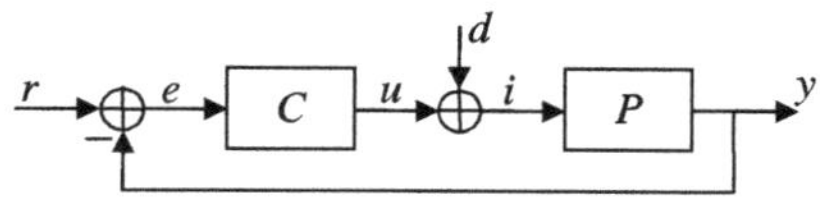

Fig. 4.1 The control scheme considered for the performance assessment and retuning methodology

The aim of the presented methodology is to evaluate the closed-loop system response when a set-point or a load disturbance step occurs and to assess the tuning of the PID controller. Then, new PID parameters are determined if the performance is not satisfactory. For the sake of simplicity, and without loss of generality, the step signal will be considered to be applied starting from null initial conditions. In order to apply the methodology, relevant process parameters have to be estimated first. For this purpose, it is worth considering the model reduction technique known as the "half rule", which states that the largest neglected (denominator) time constant is distributed evenly to the effective dead time and the smallest retained time constant [113]. In practice, the following (possibly high-order) process transfer function is considered:

$$P(s) = \frac{K}{s\prod_j (T_{j0}s+1)} e^{-L_0 s}, \tag{4.2}$$

where the time constants are ordered according to their magnitude (namely, $T_{10} > T_{20} > \cdots$). Then, a second-order integrating plus dead time (SOPDT) transfer function

$$\tilde{P}(s) = \frac{K}{s(Ts+1)} e^{-Ls} \tag{4.3}$$

is obtained by setting

$$T = T_{10} + \frac{T_{20}}{2}, \qquad L = L_0 + \frac{T_{20}}{2} + \sum_{j\geq 3} T_{i0}. \tag{4.4}$$

It is worth stressing that, by applying (4.4), it can be deduced that

$$T_0 := \sum_j T_{i0} + L_0 = T + L, \tag{4.5}$$

namely, the sum of the dead time and of the time constants of the process (4.2) is unaltered in the reduced model. Thus, T_0 is a relevant process parameter that is worth estimating for the purpose of the retuning of the PID controller, as it will be shown in the following sections.

4.3 Performance Assessment

The assessment of the performance of a control loop is generally performed by first calculating a performance index based on the available data and then by evaluating the current control performance against a selected benchmark, which represents the desired performance [45]. Usually, minimising the integrated absolute error

$$IAE = \int_0^\infty |e(t)|\,dt = \int_0^\infty |r(t) - y(t)|\,dt \tag{4.6}$$

is meaningful because this yields, in general, a low overshoot and a low settling time at the same time [110]. However, aiming at obtaining the theoretical minimum integrated absolute error that can be achieved for a single-loop system (with a general feedback controller) might not be sensible in practical cases because the robustness issue and the control effort have also to be taken into account.

If only the set-point following task is of concern, the desired performance can be selected as that obtained by applying the Internal Model Control (IMC) approach [108] which, if applied to the process (4.3) where the dead time is approximated as $e^{-Ls} = 1 - Ls$, yields a PD controller whose parameters are selected according to the following tuning rule:

$$K_p = \frac{1}{2K\lambda}, \qquad T_d = T, \tag{4.7}$$

where λ is the selected closed-loop time constant and can be selected as $\lambda = L$ according to the well-known SIMC tuning rules [113] which aim at providing a good robustness to the control system. With this PD controller and with the same approximation as before for the dead time term, the closed-loop transfer function results to be

$$\frac{Y(s)}{R(s)} = \frac{C(s)\tilde{P}(s)}{1 + C(s)\tilde{P}(s)} \cong \frac{e^{-Ls}}{1 + Ls}, \tag{4.8}$$

for which the integrated absolute error when a step signal of amplitude A_r is applied to the set-point is

$$IAE_{\mathrm{sp}} = \int_0^\infty |e(t)|\,dt = 2A_r L. \tag{4.9}$$

Thus, a sensible index, named Closed-loop Index CI, to evaluate the controller performance with respect to the set-point following task is

$$CI_{\mathrm{sp}} = \frac{2A_r L}{\int_0^\infty |e(t)|\,dt}. \tag{4.10}$$

In other words, the obtained integrated absolute error is compared with the one that would be achieved if a PD controller tuned according to the IMC tuning rules (4.7) with $\lambda = L$ is applied to the process (4.3).

In principle, the performance obtained by the control system is considered to be satisfactory if $CI_{\mathrm{sp}} = 1$. From a practical point of view, however, the controller is considered to be well tuned if $CI_{\mathrm{sp}} > \overline{CI}_{\mathrm{sp}}$ with $\overline{CI}_{\mathrm{sp}} = 0.6$. This last value has been selected by considering the (S)IMC tuning rule applied to many different processes [113], but, in any case, another value of $\overline{CI}_{\mathrm{sp}}$ can be selected by the user depending on how tight are its control specifications. Actually, selecting an empirical target value is standard practice in control loop performance assessment (see, for example, [31]).

If both the set-point following and the load disturbance rejection are of concern in the control task, the integral action has to be employed in order to achieve a null steady-state error in the presence of a constant load disturbance, and the desired performance is selected as the one obtained by applying the SIMC tuning rule [113]:

$$K_p = \frac{1}{2KL}, \qquad T_i = 8L, \qquad T_d = T. \tag{4.11}$$

In this case, the complementary sensitivity transfer function results to be

$$\frac{Y(s)}{R(s)} = \frac{C(s)\tilde{P}(s)}{1 + C(s)\tilde{P}(s)} \cong \frac{(1+8Ls)e^{-Ls}}{8L^2s^2 + 7Ls + 1}, \tag{4.12}$$

whereas, after trivial calculations, the transfer function between the load disturbance d and the process output y results to be

$$\frac{Y(s)}{D(s)} = \frac{\tilde{P}(s)}{1 + C(s)\tilde{P}(s)} \cong \frac{16L^2\mu s}{(8L^2s^2 + 7Ls + 1)(Ts+1)} e^{-Ls}. \tag{4.13}$$

By considering the transfer function (4.12), it can be determined that the integrated absolute error when a step signal of amplitude A_r is applied to the set-point can be calculated as [127]

$$IAE_{\text{sp,load}} = \int_0^\infty |e(t)|\, dt = 3.45 A_r L. \tag{4.14}$$

Thus, in this context the Closed-loop Index CI to evaluate the controller performance when a step signal is applied to the set-point can be defined as

$$CI_{\text{sp,load}} = \frac{3.45 A_r L}{\int_0^\infty |e(t)|\, dt}. \tag{4.15}$$

Conversely, when a step load disturbance of amplitude A_d occurs, the transfer function (4.13) between the load disturbance and the process output has to be considered. It is easy to see that there are no complex poles in the transfer function (4.13), and therefore there are no oscillations in the step response. Thus, the integrated error can be calculated instead of the integrated absolute error. It results to be

$$\begin{aligned} IE_{\text{load,sp}} &= \int_0^\infty e(t)\, dt = \lim_{s\to 0} s\frac{1}{s}E(s) \\ &= \lim_{s\to 0} s\frac{1}{s}\left(-\frac{\tilde{P}(s)}{1 + C(s)\tilde{P}(s)}\right) D(s) \\ &= \lim_{s\to 0} s\frac{1}{s}\left(-\frac{\tilde{P}(s)}{1 + C(s)\tilde{P}(s)}\right) \frac{A_d}{s} \end{aligned}$$

$$= -\frac{T_i}{K_p} A_d, \tag{4.16}$$

which, by taking into account (4.11), yields

$$IE_{\text{load,sp}} = -\frac{8L}{\frac{1}{2KL}} A_d = -16KL^2 A_d. \tag{4.17}$$

Hence, by taking into account that

$$IAE_{\text{load,sp}} := -IE_{\text{load,sp}} = 16KL^2 A_d, \tag{4.18}$$

the Closed-loop Index CI to evaluate the controller performance when a step load disturbance occurs can be defined as

$$CI_{\text{load,sp}} = \frac{16KL^2 A_d}{\int_0^\infty |e(t)|\, dt}. \tag{4.19}$$

Note again that both in (4.15) and (4.19) the current control performance (at the denominator) is evaluated against the target one (at the numerator). As for the set-point following performance, from a practical point of view, the controller is considered to be well tuned if $CI_{\text{sp,load}} > \overline{CI}_{\text{sp,load}}$ or if $CI_{\text{load,sp}} > \overline{CI}_{\text{load,sp}}$ with $\overline{CI}_{\text{sp,load}} = \overline{CI}_{\text{load,sp}} = 0.6$.

Finally, the case where only the load disturbance rejection performance is of concern is considered. In this case, an appropriate tuning rule, which aims at minimising the integrated absolute error, has been suggested in [110] as

$$K_p = \frac{0.78}{K(T+L)}, \qquad T_i = 1.38(T+L), \qquad T_d = 0.66(T+L). \tag{4.20}$$

With this tuning rule, it is difficult to evaluate an analytical expression for the integrated absolute error. However, a numerical procedure can be employed, yielding the following expression of the integrated absolute error:

$$IAE_{\text{load}} = A_d K T^2 \left[2.715 + 5.144\left(\frac{L}{T}\right) + 2.266\left(\frac{L}{T}\right)^2 \right], \tag{4.21}$$

and, therefore, the corresponding Closed-loop Index CI is defined as

$$CI_{\text{load}} = \frac{A_d K T^2 [2.715 + 5.144(\frac{L}{T}) + 2.266(\frac{L}{T})^2]}{\int_0^\infty |e(t)|\, dt}. \tag{4.22}$$

Also in this case the controller can be considered to be well tuned if $CI_{\text{load}} > \overline{CI}_{\text{load}}$ with $\overline{CI}_{\text{load}} = 0.6$.

It is worth stressing that the obtained value of the closed-loop index in the different cases can be also greater than one because the target values of the integrated absolute error have been chosen greater than the theoretical minimum integrated absolute error that can be achieved [39]. Indeed, the selected tuning rules [110, 113] address properly the robustness issue and the control activity issue. Note also that, in the proposed method, the occurrence of an abrupt (namely, step-like) load disturbance has been assumed. Indeed, this is the most relevant case for the control system, as the disturbance excites significantly the dynamics of the control system itself [110]. Thus, the performance assessment technique has to be implemented together with a procedure for the detection of abrupt load disturbances. Methods for this purpose have been proposed in [32, 125].

4.4 Estimation of the Process Parameters

For the computation of the Closed-loop Index CI and, as it will be clear in Section 4.5, for the retuning of the PID controller (see (4.7), (4.11), and (4.20)), it is necessary to estimate the values of K, T_0 and the value of the apparent dead time L, as well as the value of the amplitude of the load step disturbance A_d (the value of A_r is obviously known). For this purpose, the closed-loop response of either a set-point step or a load disturbance step signal can be evaluated effectively. The two cases are considered separately hereafter.

4.4.1 Set-point Step Response

In case a step signal of amplitude A_r is applied to the set-point, the apparent dead time L of the system can be evaluated by considering the time interval from the occurrence of a set-point step signal and the time instant when the process output attains the 2% of the new set-point value A_r, namely, when the condition $y > 0.02A_r$ occurs. Actually, from a practical point of view, in order to cope with the measurement noise, a simple sensible solution is to define a noise band NB [8] (whose amplitude should be equal to the amplitude of the measurement noise) and to rewrite the condition as $y > NB$ (see Section 2.2.1.1).

The determination of the sum of the lags and of the dead time of the process can be performed by evaluating the control system response as well. In particular, the following variable can be considered:

$$e_u(t) := K \int_0^t u(v)\,dv - y(t). \tag{4.23}$$

By applying the Laplace transform to (4.23) and by expressing u and y in terms of r (whose Laplace transform is $R(s) = A_r/s$), it is

$$\begin{aligned} E_u(s) &= K\frac{1}{s}U(s) - Y(s) \\ &= K\frac{1}{s}\frac{C(s)}{1+C(s)P(s)}R(s) - \frac{C(s)P(s)}{1+C(s)P(s)}R(s) \end{aligned}$$

$$= \frac{A_r}{s^2} \frac{C(s)(K - sP(s))}{1 + C(s)P(s)}. \tag{4.24}$$

At this point, for the sake of clarity, it is convenient to write the controller and process transfer functions respectively as

$$C(s) = \frac{K_p}{T_i s} \tilde{c}(s), \quad \text{where } \tilde{c}(s) := (T_i s + 1)(T_d s + 1), \tag{4.25}$$

and

$$P(s) = \frac{K e^{-L_0 s}}{s q(s)}, \quad \text{where } q(s) := \prod_j (T_{j0} s + 1). \tag{4.26}$$

Then, Expression (4.24) can be rewritten as

$$E_u(s) = \frac{A_r}{s} \frac{K K_p \tilde{c}(s)}{T_i s^2 q(s) + K K_p \tilde{c}(s) e^{-L_0 s}} \left(q(s) - e^{-L_0 s}\right). \tag{4.27}$$

By applying the final value theorem to the integral of e_u it is eventually obtained that

$$\begin{aligned} \lim_{t\to+\infty} \int_0^t e_u(v)\,dv &= \lim_{s\to 0} s \frac{1}{s} A_r \frac{K K_p \tilde{c}(s)}{T_i s^2 q(s) + K K_p \tilde{c}(s) e^{-L_0 s}} \frac{q(s) - e^{-L_0 s}}{s} \\ &= A_r \lim_{s\to 0} \left\{ \frac{1 - e^{-L_0 s}}{s} + \frac{q(s) - 1}{s} \right\} \\ &= A_r \left(L_0 + \sum_j T_{j0} \right) \\ &= A_r T_0. \end{aligned} \tag{4.28}$$

Thus, the sum of the lags and of the dead time of the process can be obtained by evaluating the integral of $e_u(t)$ at the steady-state (which does not depend on the PID parameters) when a step signal is applied to the set-point and by dividing it by the amplitude A_r of the step.

The process gain K can be determined by twice integrating the control error. In fact, the Laplace transform of the control error can be expressed as

$$E(s) = \frac{A_r}{s} \frac{1}{1 + C(s)P(s)} = \frac{A_r s q(s) T_i}{T_i s^2 q(s) + K K_p \tilde{c}(s) e^{-L_0 s}}, \tag{4.29}$$

and, therefore, it can be deduced that

$$\lim_{t\to+\infty} \int_0^t \int_0^{v_2} e(v_1)\,dv_1\,dv_2 = \lim_{s\to 0} s \frac{1}{s^2} \frac{A_r s q(s) T_i}{T_i s^2 q(s) + K K_p \tilde{c}(s) e^{-L_0 s}} = \frac{A_r T_i}{\mu K_p}. \tag{4.30}$$

Thus, the gain of the process can be determined easily as

$$K = A_r \frac{T_i}{K_p \int_0^\infty \int_0^t e(v)\, dv\, dt}. \tag{4.31}$$

In case the integral action is not employed, namely $C(s) = K_p(T_d s + 1)$, a similar reasoning can be applied, yielding

$$E(s) = \frac{A_r}{s} \frac{1}{1 + C(s)P(s)} = \frac{A_r q(s)}{sq(s) + KK_p(1 + T_d s)e^{-L_0 s}}, \tag{4.32}$$

and, therefore,

$$\lim_{t \to +\infty} \int_0^t e(v)\, dv = \lim_{s \to 0} s \frac{1}{s} \frac{A_r q(s)}{sq(s) + KK_p(1 + T_d s)e^{-L_0 s}} = \frac{A_r}{KK_p}, \tag{4.33}$$

which yields

$$K = \frac{A_r}{K_p \int_0^\infty e(t)\, dt}. \tag{4.34}$$

It is worth noting that both the value of the gain and of the sum of the lags and of the dead time of the process are determined by considering the integral of signals and therefore the effect of the measurement noise is reduced [117].

4.4.2 Load Disturbance Step Response

The determination of the process parameters can also be performed by evaluating the control system response to a step load disturbance, provided that the instant when the disturbance occurs can be measured (for the sake of simplicity, it is assumed that the load disturbance occurs at time $t = 0$). In this context, the apparent dead time of the process can be determined by applying a reasoning similar to the set-point step case (namely, by exploiting the noise band concept).

Then, the amplitude A_d of the step load disturbance can be determined by considering the final value of the integral of the control error. In fact, the expression of the Laplace transform of the control error is

$$E(s) = -\frac{P(s)}{1 + C(s)P(s)} D(s) = -\frac{T_i s K e^{-L_0}}{T_i s^2 q(s) + K_p \tilde{c}(s) K e^{-L_0 s}} \frac{A_d}{s}, \tag{4.35}$$

and, therefore,

$$\lim_{t \to +\infty} \int_0^t e(v)\, dv = \lim_{s \to 0} s \frac{1}{s} \frac{A_d}{s} \left(-\frac{T_i s K e^{-L_0}}{T_i s q(s) + K_p \tilde{c}(s) K e^{-L_0 s}} \right) = -\frac{A_d T_i}{K_p}. \tag{4.36}$$

Thus, the amplitude of the step disturbance can be determined as

$$A_d = -\frac{K_p}{T_i}\int_0^{\infty} e(v)\,dv. \tag{4.37}$$

Once the amplitude of the step disturbance has been determined, the process gain K can be determined by first considering the Laplace transform of the process input $i = u + d$, that is,

$$\begin{aligned} I(s) &= U(s) + D(s) \\ &= -\frac{C(s)P(s)}{1 + C(s)P(s)}D(s) + D(s) \\ &= \frac{T_i s^2 q(s)}{T_i s^2 q(s) + K_p \tilde{c}(s) K e^{-L_0 s}}\frac{A_d}{s}. \end{aligned} \tag{4.38}$$

Thus, if $i(t)$ is integrated twice, the limit as $t \to +\infty$ is determined as

$$\lim_{t\to+\infty}\int_0^t\int_0^{v_2} i(v_1)\,dv_1\,dv_2 = \lim_{s\to 0} s\frac{1}{s^2}\frac{T_i s^2 q(s)}{T_i s^2 q(s) + K_p\tilde{c}(s)Ke^{-L_0 s}}\frac{A_d}{s} = \frac{T_i A_d}{K K_p}. \tag{4.39}$$

The process gain K can be therefore found easily, once the value of A_d has been determined, by using (4.37), as

$$K = A_d\frac{T_i}{K_p\int_0^{\infty}\int_0^t (u(v) + A_d)\,dv\,dt}. \tag{4.40}$$

Finally, the determination of the sum of the lags and of the dead time of the process can be performed by initially considering the variable

$$e_i(t) := K\int_0^t i(v)\,dv - y(t). \tag{4.41}$$

By applying the Laplace transform to (4.41) and by expressing u and y in terms of d, it can be deduced that

$$\begin{aligned} E_i(s) &= K\frac{1}{s}\big(U(s) + D(s)\big) - Y(s) \\ &= K\frac{1}{s}\left[-\frac{C(s)P(s)}{1 + C(s)P(s)}D(s) + D(s)\right] - \frac{P(s)}{1 + C(s)P(s)}D(s) \\ &= \frac{KT_i A_d(q(s) - e^{-L_0 s})}{T_i s^2 q(s) + K_p K\tilde{c}(s)e^{-L_0 s}}. \end{aligned} \tag{4.42}$$

By integrating twice e_i and by applying the final value theorem, it is (see (4.28))

$$\lim_{t\to+\infty}\int_0^t\int_0^{v_2} e_i(v_1)\,dv_1\,dv_2 = \lim_{s\to 0} s\frac{1}{s}\frac{KT_i A_d}{T_i s^2 q(s) + KK_p\tilde{c}(s)e^{-L_0 s}}\frac{q(s) - e^{-L_0 s}}{s}$$

$$= \frac{T_i A_d}{K_p} T_0. \tag{4.43}$$

Thus, T_0 can be obtained as

$$T_0 = \frac{K_p}{T_i A_d} \int_0^{\infty} \int_0^{t} e_i(v)\, dv\, dt. \tag{4.44}$$

For the purpose of determining the closed-loop index (4.22), the time constant T can be calculated as $T = T_0 - L$.

4.5 Retuning of the PID Controller

If the performance provided by the controller results to be unsatisfactory (see Section 4.3), the PID controller needs to be retuned according to the required control task. In particular, if only the set-point following is of concern, the retuning procedure aims at achieving the same performance of the IMC tuning rules (4.7). For this purpose, once the process parameters have been estimated according to the technique described in Section 4.4, the PD controller parameters are selected as

$$K_p = \frac{1}{2KL}, \qquad T_d = T_0 - L. \tag{4.45}$$

When both the set-point following and the load disturbance rejection performance have to be addressed, the target tuning rule is the SIMC one (see (4.7)), and this can be achieved by selecting

$$K_p = \frac{1}{2KL}, \qquad T_i = 8L, \qquad T_d = T_0 - L. \tag{4.46}$$

Finally, when only the load disturbance rejection performance is of concern, the tuning rule (4.20) can be pursued by selecting (note that in this case the apparent dead time estimation is not required)

$$K_p = \frac{0.78}{KT_0}, \qquad T_i = 1.38T_0, \qquad T_d = 0.66T_0. \tag{4.47}$$

It is worth noting that, although, in principle, the devised methodology exploits set-point changes and load disturbances that naturally occur in the process operations, it can be also obviously employed in an automatic (one shot) tuning context, where the retuning of the PID parameters is requested explicitly by the operator, who applies the step input signal on purpose.

4.6 Simulation Results

Some simulation results are presented hereafter in order to evaluate the effectiveness of the presented method in different contexts.

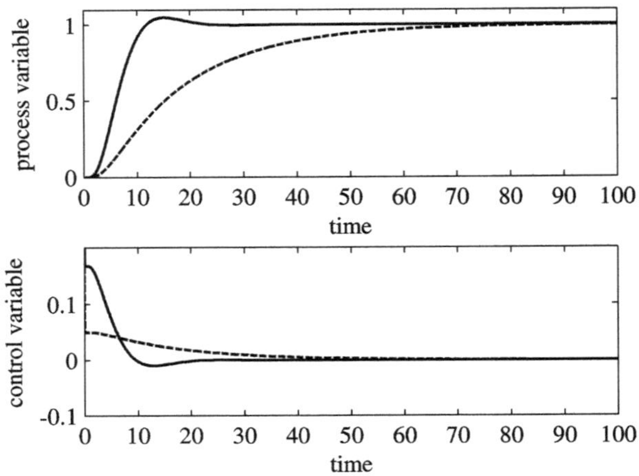

Fig. 4.2 Step response in Example 1. *Dashed line*: PD with the initial tuning $K_p = 0.05$ and $T_d = 1$. *Solid line*: PD with the new tuning $K_p = 0.17$ and $T_d = 1.5$

4.6.1 Example 1

Consider the process

$$P(s) = \frac{1}{s(s+1)^4} e^{-0.5s}. \tag{4.48}$$

As a first case, only the performance related to the set-point step response is considered. If a PD controller is initially tuned with $K_p = 0.05$ and $T_d = 1$, the closed-loop unit step response shown in Figure 4.2 as a dashed line is obtained. The resulting performance index is $CI_{\text{sp}} = 0.3$, and the resulting integrated absolute error and 2% settling time are respectively $IAE = 19.96$ and $T_{s2} = 66.62$. By applying the retuning procedure, the process parameters are estimated as $L = 2.98$, $T_0 = 4.49$, and $K = 1$ (see Section 4.4.1), and the new PD parameters are determined from (4.45) as $K_p = 0.17$ and $T_d = 1.5$. The corresponding unit step response is shown in Figure 4.2 as a solid line. The resulting performance index is $CI_{\text{sp}} = 0.6$, and the resulting integrated absolute error and 2% settling time are respectively $IAE = 6.65$ and $T_{s2} = 20.16$.

Conversely, if a PI controller is initially employed with $K_p = 0.2$ and $T_i = 10$, the (very oscillatory) step response shown in Figure 4.3 as a dashed line is obtained, with $CI_{\text{sp}} = 0.17$, $IAE = 31.51$, and $T_{s2} = 136.77$. By applying the retuning procedure the process parameters are estimated as $L = 2.63$, $T_0 = 4.49$, and $K = 1$, and from (4.45) the parameters of the PD controller are determined as $K_p = 0.19$ and $T_d = 1.87$. The corresponding step response is shown in Figure 4.3 as a solid line with $CI_{\text{sp}} = 0.63$, $IAE = 5.89$, and $T_{s2} = 17.11$. Note that the slight difference with the previous case is due to the slight difference of the estimated values of the dead time.

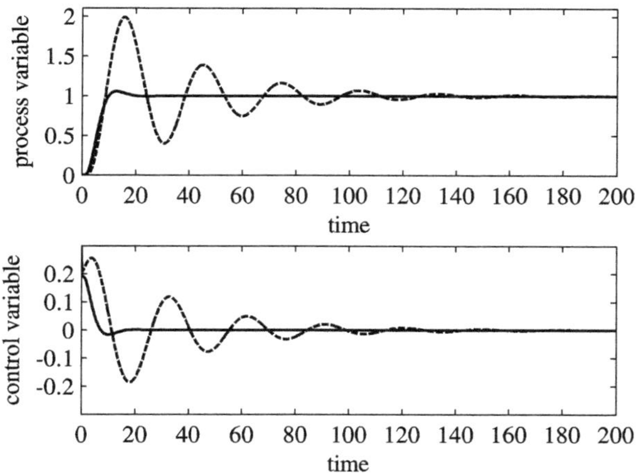

Fig. 4.3 Step response in Example 1. *Dashed line*: PI with the initial tuning $K_p = 0.2$ and $T_i = 10$. *Solid line*: PD with the new tuning $K_p = 0.19$ and $T_d = 1.87$

4.6.2 Example 2

Process (4.48) is considered again, but in this case both the set-point following and the load disturbance rejection performance are of concern. The PID controller parameters are initially selected as $K_p = 0.05$, $T_i = 50$, and $T_d = 4$. By evaluating the set-point step response, the process parameters are estimated as $L = 2.21$, $T_0 = 4.5$, and $K = 1$ (see Section 4.4.1). Then, the PID controller is retuned according to (4.46) as $K_p = 0.22$, $T_i = 17.73$, and $T_d = 2.28$. The process variable in the two cases where a unit step is applied to the set-point signal at $t = 0$ and a step load disturbance of amplitude $A_d = 0.1$ occurs at time $t = 250$ is shown in Figure 4.4. Regarding the set-point following task, the performance index changes from $CI_{\text{sp,load}} = 0.28$ to $CI_{\text{sp,load}} = 0.66$, whereas the integrated absolute error changes from $IAE = 0.28$ to $IAE = 0.09$ and the 2% settling time from $T_{s2} = 158.1$ to $T_{s2} = 43.5$. Regarding the load disturbance rejection task, the performance index changes from $CI_{\text{load,sp}} = 0.06$ to $CI_{\text{load,sp}} = 0.74$, whereas the integrated absolute error changes from $IAE = 103.3$ to $IAE = 8.06$ and the 2% settling time from $T_{s2} = 216.4$ to $T_{s2} = 50.84$.

Conversely, if the load disturbance response is initially evaluated (again with $K_p = 0.05$, $T_i = 50$, and $T_d = 4$), the process parameters are estimated as $L = 1.93$, $A_d = 0.1$, $K = 1$, and $T_0 = 4.5$, which yields the new values of the parameters $K_p = 0.26$, $T_i = 15.47$, and $T_d = 2.57$. The closed-loop response to a set-point and a load disturbance step is shown in Figure 4.5 (where the initial case is also shown again for the sake of comparison). Regarding the load disturbance rejection task, the new performance index is $CI_{\text{load,sp}} = 1$, the integrated absolute error is $IAE = 5.98$, and the 2% settling time is $T_{s2} = 75.3$. Regarding the set-point following task, the new performance index is $CI_{\text{load,sp}} = 0.70$, the new integrated absolute error is $IAE = 8.13$, and the 2% settling time is $T_{s2} = 38.65$.

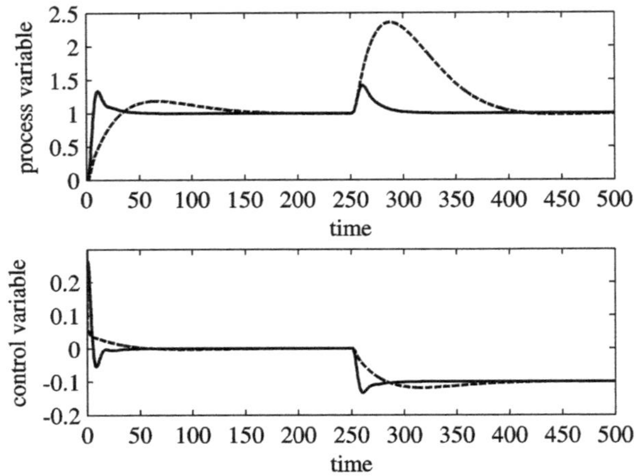

Fig. 4.4 Step response in Example 2. *Dashed line*: PID with the initial tuning $K_p = 0.05$, $T_i = 50$, and $T_d = 4$. *Solid line*: PID with the new tuning $K_p = 0.22$, $T_i = 17.73$, and $T_d = 2.28$

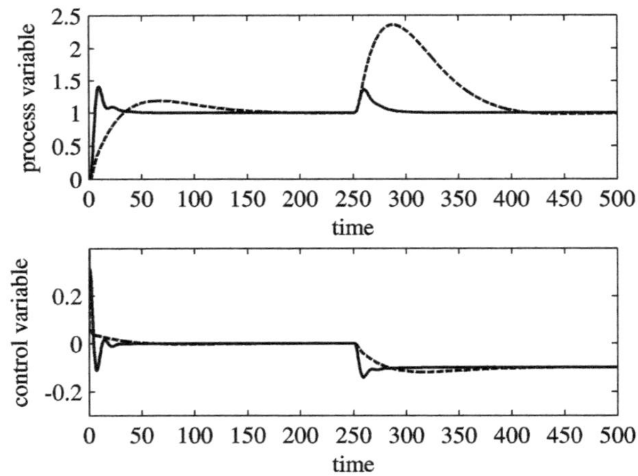

Fig. 4.5 Step response in Example 2. *Dashed line*: PID with the initial tuning $K_p = 0.05$, $T_i = 50$, and $T_d = 4$. *Solid line*: PID with the new tuning $K_p = 0.26$, $T_i = 15.47$, and $T_d = 2.57$

4.6.3 Example 3

Process (4.48) is considered again, but in this case only the load disturbance rejection performance is addressed. If initially the PID controller parameters are $K_p = 0.05$, $T_i = 50$, and $T_d = 4$ and the load disturbance step response ($A_d = 0.1$) is evaluated, the process parameters are estimated as $L = 1.93$, $A_d = 0.1$, $K = 1$, and $T_0 = 4.5$ (see Section 4.4.2). Then, the PID controller is retuned according to (4.47) as $K_p = 0.17$, $T_i = 6.21$, and $T_d = 2.97$. The process variable in the two cases where a step signal of amplitude $A_d = 0.1$ is added to the control variable signal at

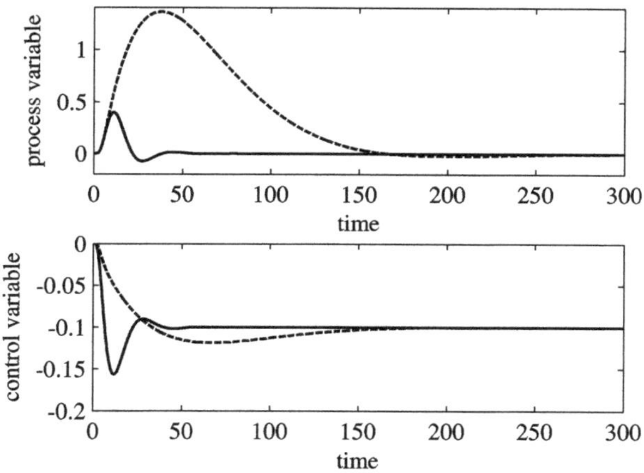

Fig. 4.6 Load disturbance step response in Example 3. *Dashed line*: PID with the initial tuning $K_p = 0.05$, $T_i = 50$, and $T_d = 4$. *Solid line*: PID with the new tuning $K_p = 0.17$, $T_i = 6.21$, and $T_d = 2.97$

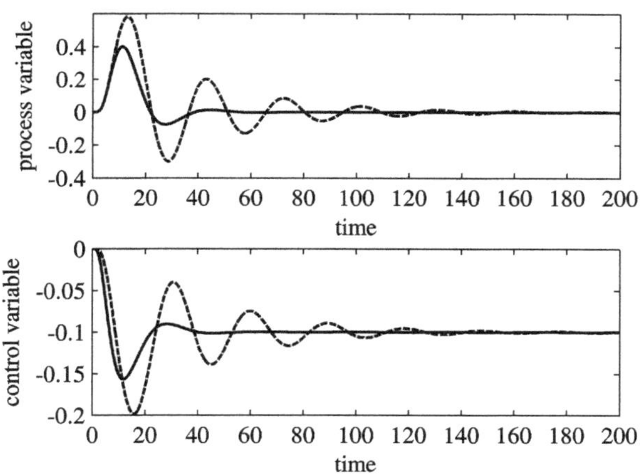

Fig. 4.7 Load disturbance step response in Example 3. *Dashed line*: PID with the initial tuning $K_p = 0.2$, $T_i = 10$, and $T_d = 0$. *Solid line*: PID with the new tuning $K_p = 0.17$, $T_i = 6.21$, and $T_d = 2.97$

$t = 0$ is shown in Figure 4.6. The performance index improves from $CI_{\text{load}} = 0.05$ to $CI_{\text{load}} = 1$, whereas the integrated absolute error decreases from $IAE = 103.3$ to $IAE = 5.15$ and the settling time decreases from $T_{s2} = 216.6$ to $T_{s2} = 35.77$.

If the initial settings are changed to $K_p = 0.2$, $T_i = 10$, and $T_d = 0$, the resulting performance indexes are $CI_{\text{load}} = 0.36$, $IAE = 14.51$, $T_{s2} = 118.5$, the process parameters are estimated as $L = 1.93$, $A_d = 0.1$, $K = 1$, and $T_0 = 4.5$, and the PID gains are modified again as $K_p = 0.17$, $T_i = 6.21$, and $T_d = 2.97$. The same performance indexes of the previous case are found. The resulting process variables are plotted in Figure 4.7.

4.7 Conclusions

In this chapter a methodology for the performance assessment and retuning of PID controllers has been described. The main feature of this methodology is that the performance of the (pre-existing) controller is evaluated and that the new PID controller parameters are selected without the need of special experiments, but just considering (abrupt) set-point changes or load disturbances. Different control specifications can be addressed.

Chapter 5
Plug&Control

One of the main reasons for the success of PID controllers in the industrial context is their relative ease of use. Indeed, the fast commissioning of the controller is essential in many applications, where a tight performance is not required, in order to reduce the implementation costs. In this context, the availability of the so-called Plug&Control function (*i.e.*, to automatically make the controller work properly after simply connecting it in the control architecture, without further intervention from the operator) is highly desirable.

With respect to the classic automatic tuning procedures, this function has the advantage that a dedicated identification (possibly time-consuming) experiment is not required, since the estimation of the process parameters is performed during the normal start-up of the process. This might allow a significant saving of time, energy, and material.

A methodology related to this topic is described in this chapter, where only the case of IPDT processes is addressed (for self-regulating processes, see [93, 129, 132, 133]).

5.1 Methodology

A time-optimal Plug&Control strategy for IPDT processes has been first proposed in [129]. It is based on the combined use of three-state and PID control to perform a transition from one set-point value to another, as required by the process start-up operation, and it is applicable without a priori knowledge of the process model parameters, with the exception of the sign of the process gain, which will be assumed to be positive from now on without loss of generality.

Basically, the methodology consists of initially setting the controller output at its upper limit when the step on the set-point signal is applied. Afterwards, when the process output leaves its previous value, the dead time L of the process is estimated. Then, from this instant the process parameters are estimated through a least squares procedure. Once the process model is estimated, a time-optimal control strategy, based on the saturation limits of the actuator, can be computed and applied. At the

A. Visioli, Q.-C. Zhong, *Control of Integral Processes with Dead Time*, Advances in Industrial Control,
DOI 10.1007/978-0-85729-070-0_5, © Springer-Verlag London Limited 2011

same time, the PID controller (which is not adopted in this phase) can be properly tuned according to a selected tuning rule. If the process parameters are perfectly estimated, then at the end of the time-optimal control, the process output would be exactly at its desired steady-state value. However, estimation inaccuracies are not avoidable in practical cases, mainly due to the presence of measurement noise and numerical approximations. Therefore, at the end of the time-optimal strategy, when the process output is actually close to its desired value, the controller is set to the PID mode. In this way, the desired output value is actually attained, and possible subsequent load disturbances can be compensated.

5.2 Algorithm

Consider the IPDT process

$$P(s) = \frac{K}{s}e^{-Ls}, \quad K > 0,\ T > 0. \tag{5.1}$$

Denote u as the process input, and y as the process output. Suppose now that an output transition from y_0 to $y_{sp} = y_0 + y_1$ is then required to be performed, starting from time t_0 (assume that the process is at an equilibrium point with $u_0 := 0$ and $y_0 := y(t_0)$). For the sake of simplicity and without loss of generality, hereafter it will be assumed that $y_1 > 0$.

The following time-optimal Plug&Control (TOPC) strategy can be applied. For the sake of clarity, the algorithm presented is related to the noise-free case. Modifications to be carried out in order to cope with noise and other practical issues are discussed in Section 5.3.

TOPC algorithm

1. Set u_{max} and u_{min} as the maximum and minimum values respectively of the control variable u during the three-state control and calculate $u^+ = u_{max} - u_0$ and $u^- = u_{min} - u_0$.
2. Set flag $= 1$.
3. At time $t = t_0$ set $u = u_{max}$.
4. When $y > y_0$, set $t_1 = t$ and $\hat{L} = t_1 - t_0$ (estimated dead time of the process).
5. At time $t = t_1$ start a recursive least squares algorithm [7, page 51].
6. When $|\hat{K}(t) - \hat{K}(t - \Delta t)| < \varepsilon$ ($\hat{K}$ is the estimated gain of the process):
   (a) Set $t_2 = t$.
   (b) Set $\hat{K} = \hat{K}(t_2)$.
   (c) Apply a PI(D) tuning rule based on the model identified.
   (d) Calculate
$$t_{s1} = t_0 + \frac{y_1}{\hat{K}u^+}. \tag{5.2}$$
   (e) If $t_{s1} < t_2$, then set $t_{s1} = t_2$, flag $= 0$ and calculate
$$t_{s2} = -\frac{u^+(t_{s1} - t_0) - u^-(t_{s1} - t_0) - \frac{y_1}{\hat{K}}}{u^-}. \tag{5.3}$$

7. If flag $= 1$, then set $u = u_{\max}$ when $t \leq t_{s1}$ and $u = 0$ when $t > t_{s1}$, else set $u = u_{\min}$ when $t \leq t_{s2}$ and $u = 0$ when $t > t_{s2}$.
8. When $t > \hat{L} + t_{s1}$ (if flag $= 1$) or when $t > \hat{L} + t_{s2}$ (if flag $= 0$), apply the PI(D) controller.

It can be seen that the algorithm requires that, when a set-point change is required at time $t = t_0$, the control variable is set to its maximum level $u = u_{\max}$. Then, when the process output leaves its initial value y_0 at time $t = t_1$, the dead time L of the process is detected. A standard recursive least-squares algorithm is then applied. When the estimation of the parameter K converges (*i.e.*, when the difference of two successive estimations is less than a predefined threshold ε) at time $t = t_2$, a model of the process is available. This can be exploited to tune the PID controller and to determine a time-optimal strategy to attain the set-point value. In particular, the time interval for which the control variable has to be kept at its maximum value $u_{\max}$ in order for the process output to attain the set-point value in the minimum time, namely, the time instant t_{s1} when the value of the control variable has to be switched from $u_{\max}$ to the final null steady-state value is determined (Equation (5.2) can be trivially derived in the context of the optimal control theory [57]).

It may happen that, because of the large dead time of the process or because of the large time interval required for the identification procedure to converge, it follows that $t_{s1} < t_2$, *i.e.*, that the control variable has been set to its maximum value for a larger time than requested by the time-optimal control (this condition determines the setting flag $= 0$). This means that the output, even in the perfect match case (*i.e.*, even if the process parameters are perfectly estimated) presents an overshoot. Hence, the control variable must be set immediately at its minimum level and kept at this value for a determined time interval $t_{s2} - t_{s1}$ (the switching time t_{s2} is also derived in the context of optimal theory). Then, the control variable is set at the null steady-state value. At the end of the (three-state) time-optimal control strategy, if the process model is perfectly estimated, the process output attains its set-point value.

As in practical cases this does not occur because of the unavoidable estimation inaccuracies, the PID controller is applied to cancel the remaining steady-state error and to cope with subsequent possible load disturbances.

5.3 Practical Considerations

A few technical problems have to be solved in order to effectively apply the TOPC algorithm in practical cases. First, as real measurements are always corrupted with noise, the condition $y > y_0$ at step 4 should be substituted with $y > y_0 + NB$, where *NB* is the estimated noise band [8], namely, a threshold value that determines whether the process variable has started increasing. Specifically, if $y < y_0 + NB$, then y is considered to be equal to y_0. The value of *NB* can be easily selected by monitoring that process output for a sufficiently long time when the process is at steady state.

It has also to be noted that it is not strictly necessary for the control constraints $u_{\min}$ and $u_{\max}$ to correspond to the actual physical limits of the actuator. Actually, more conservative bounds can be selected for various operating reasons or to preserve the linearity of the model.

Then, the recursive least squares algorithm in step 5 has to be initialised. This can be easily done by selecting a very rough estimate of the process gain denoted as $\hat{K}_0$ (see Section 5.4). The value of ε has to be fixed as well. Actually, by fixing it at a low value the user is confident that the identification phase is ended with satisfactory accuracy.

Then, a bumpless transfer [5] has to be applied at step 8 at the time of switching from the three-state to the PID controller.

Finally, it has to be highlighted that the proposed method could be applied in the context of feedforward control of set-point steps with "full power" plus PID control [93]. In other words, for sufficiently large set-point steps, a control zone has to be defined: inside a narrow band around the set-point, a closed-loop PID control is employed, while outside the control zone, the controller output is set at its maximum value. Note that with such a strategy, the integral time constant of the PID controller can be increased without the occurrence of overshoots and yielding at the same time a faster load disturbance rejection. The identification procedure can be skipped if the process parameters have already been estimated, or it can be applied if variations of the process parameters are detected.

5.4 Simulation Results

In order to understand better the TOPC algorithm and to verify its effectiveness, a few simulation results are presented hereafter. The following process is considered:

$$P_1(s) = \frac{0.1}{s} e^{-s}. \tag{5.4}$$

The initial conditions are fixed to $t_0 = 0$, $y_0 = 0$, and $u_0 = 0$. Then, it is set $y_{sp} = y_1 = 1$, $u_{max} = 1.5$, and $u_{min} = -1.5$. Here, a noise band of $NB = 0.01$ is set. The recursive least squares procedure parameters are $\hat{K}_0 = 0.5$ and $\varepsilon = 10^{-3}$. The tuning formula employed for the PI controller is [110]

$$\begin{aligned} K_p &= \frac{0.9259}{\hat{K}\hat{L}}, \\ T_i &= 4\hat{L}. \end{aligned} \tag{5.5}$$

The result obtained by applying the TOPC strategy is shown in Figure 5.1.

It is $t_1 = \hat{L} = 1.06$ and $t_2 = 2.33$ (the estimated process parameter is $\hat{K} = 0.1$). By means of Expression (5.2) the optimal switching time is determined as $t_{s1} = 6.65$, where the control variable is set to zero. Then, at time $t = t_{s1} + \hat{L} = 7.71$, the PI controller is applied ($K_p = 8.71$ and $T_i = 4.24$), and its performance in the load disturbance rejection task is evaluated at time $t = 15$. It appears that a time optimal transition is achieved, and, at the same time, the PI controller is tuned satisfactorily.

As another illustrative example, the process

$$P_2(s) = \frac{1}{s} e^{-s} \tag{5.6}$$

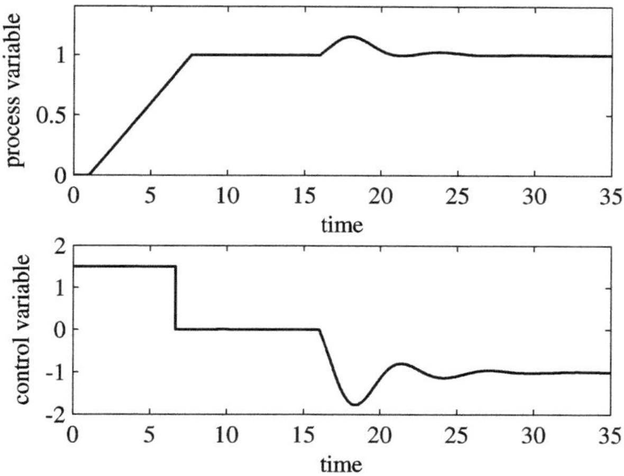

Fig. 5.1 Result of the application of the TOPC strategy with the IPDT process $P_1(s)$

is considered (note that the value of the dead time is significant with respect to the time constant). The same design parameters of the previous case are employed. The result obtained is shown in Figure 5.2. It is $t_1 = \hat{L} = 1.0$ and $t_2 = 1.13$ (the estimated process parameter is $\hat{K} = 1.012$). The optimal switching time is calculated as $t_{s1} = 0.66$. Since $t_{s1} < t_2$, t_{s1} is fixed equal to t_2, and then the optimal switching time t_{s2} is calculated by means of Expression (5.3). It results $t_{s2} = 1.601$. Thus, the control variable is set to zero for $t_{s2} < t < t_{s2} + \hat{L}$ before applying the PI controller, whose parameters have been set to $K_p = 0.915$ and $T_i = 4$.

It is worth underlying that the resulting overshoot is due to the high value of the dead time of the process, for which the control variable is set to its maximum value $u_{\max}$ for a long time interval.

From the results presented it appears that the time-optimal Plug&Control strategy is effective in providing a fast commissioning of the control loop when a tight performance is not required. Indeed, the technique is suitable for those processes where the dominant dynamics is not of high order and where possibly somewhat large overshoots are allowed (at least in the start-up phase of the process). It is worth stressing, however, that by a suitable choice of the design parameters (namely, the maximum and minimum level of the control variable during the three-state control phase) the overshoot can be significantly reduced (at the expense of the rise time). In fact, the design parameters have a clear physical meaning, and technical problems can be solved in a practical context by exploiting a reasonable knowledge of the plant.

Finally, it has to be noted that, instead of the recursive least squares algorithm, a batch least squares algorithm [117] can be applied for the identification purpose [129]. Although in this case the methodology has the disadvantage that the user has to select the part of the transient for which data are collected for the estimation of the parameters, this choice somehow allows the handling of the trade-off between

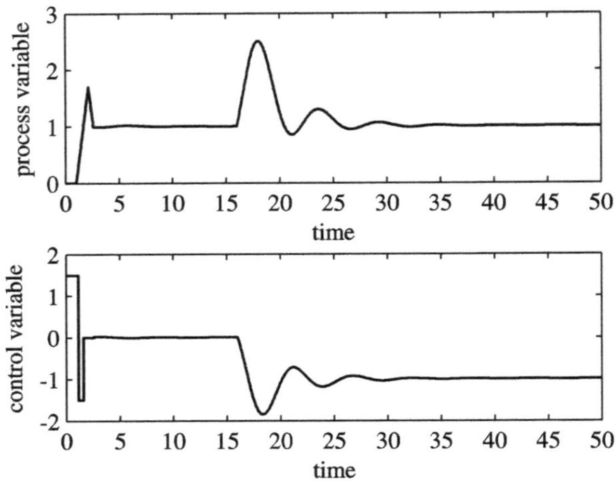

Fig. 5.2 Result of the application of the TOPC strategy with the IPDT process $P_2(s)$

estimation accuracy and the resulting overshoot, and for this reason, can be preferred by an experienced user.

5.5 Conclusions

In this chapter a Plug&Control strategy has been presented. It has been shown that this approach is very promising because, provided that the methodology is applied in a suitable context, it is capable of providing a fast and effective design of the control loop. Obviously, this feature is more relevant in large plants, when there are many (simple) loops to tune.

Part II
Two-degree-of-freedom Control Schemes

Chapter 6
Feedforward Control

The main purpose of using feedback is to compensate for external disturbances and for model uncertainties. Actually, when a sufficiently accurate model of the integral process is available (and the process dynamics does not change significantly during the process operations), control performance can be improved in general by conveniently employing an additional feedforward (open-loop) control law, thus employing a two-degree-of-freedom control. After having presented the standard two-degree-of-freedom control scheme, this chapter focuses on different methodologies for the design and the implementation of a feedforward control law, to be adopted in conjunction with the feedback action provided by a PID controller. It is shown that the problem can be approached from different points of view. In particular, regarding the set-point following task, two kinds of approaches are presented: the design of a causal feedforward action and of a noncausal feedforward action. In the first case a (nonlinear) two-state control law is described. In the second case, to be employed when desired process output transitions are known in advance, strategies based on input–output inversion are explained both in the continuous-time and discrete-time frameworks.

6.1 Standard Two-degree-of-freedom Control Scheme

As already mentioned in Section 2.1.2, it is difficult to obtain a satisfactory performance in both the load disturbance rejection and set-point following task. In fact, if the feedback controller is properly designed for the compensation of modelling uncertainties and external disturbances, an oscillatory set-point step response occurs in general. The natural solution for this problem is to employ a two-degree-of-freedom control scheme, for which the use of a set-point weighted PID controller represents a particular case. In general, the standard two-degree-of-freedom control scheme is that shown in Figure 2.3, where C is obviously the feedback controller, and F is a general feedforward controller which aims at recovering the set-point following performance. Usually, while C is designed in order to have a high gain at the gain cross-over frequency, F is designed as a low-pass filter in order to smooth the step

A. Visioli, Q.-C. Zhong, *Control of Integral Processes with Dead Time*,
Advances in Industrial Control,
DOI 10.1007/978-0-85729-070-0_6, © Springer-Verlag London Limited 2011

signal and therefore to reduce the overshoot and the oscillations in the set-point response. However, as a drawback, the rise time increases. In the following sections it will be shown that a different choice of the feedforward action can improve the performance significantly by reducing the oscillations and the rise time at the same time and by considering the actuator constraints as well.

6.2 Two-state Time-optimal Feedforward Control

In the framework of causal feedforward control actions, an effective technique is based on the use of a bang-bang type control, as it is explained hereafter.

6.2.1 *Methodology*

A significant improvement in the set-point following performances can be obtained by employing a two-state (nonlinear) feedforward control law, as shown in [130, 135, 136], where a technique inspired by the bang-bang control strategy [57] is devised to achieve a fast response to set-point changes. With the aim of fully exploiting the capabilities of the actuator, namely, in order to take into account the actuator nonlinearity (without impairing the ease of use of the overall control system), the following methodology can be employed.

Assume that it is required to design a control scheme based on a PID controller plus a feedforward term aimed at achieving a transition of the process output y from the value y_0 to the value y_1 in a predefined time interval of duration τ. Hereafter, for the sake of clarity and without loss of generality, it will be assumed that $y_0 = 0$ and $y_1 > 0$.

The devised PID plus feedforward control scheme is shown in Figure 6.1 and implements the following design technique (note that the feedforward control action is added to the controller output but the devised scheme can be trivially modified in order for the feedforward control action to be applied to the closed-loop system as in Figure 2.3). First, the process is described by a IPDT model, *i.e.*,

$$P(s) = \frac{K}{s} e^{-Ls}. \tag{6.1}$$

Based on this model, the output u_{ff} of the feedforward block FF is defined as follows:

$$u_{\mathrm{ff}}(t) = \begin{cases} \bar{u}_{\mathrm{ff}} & \text{if } t < \tau, \\ 0 & \text{if } t \geq \tau, \end{cases} \tag{6.2}$$

where the value of $\bar{u}_{\mathrm{ff}}$ is determined, after trivial calculations, in such a way that the process output y (which is necessarily zero until time $t = L$) is y_1 at time $t = \tau + L$.

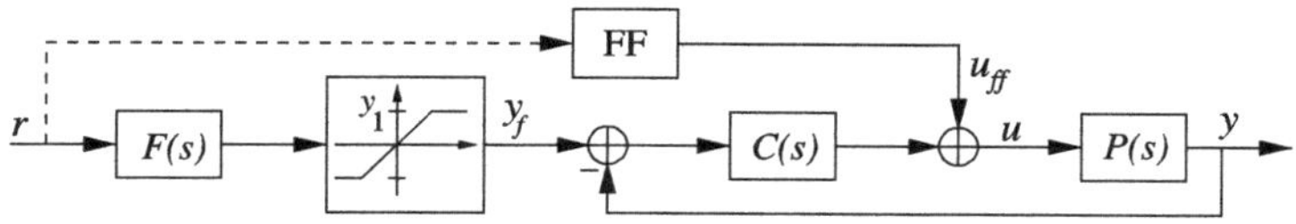

Fig. 6.1 Block diagram of the PID plus nonlinear feedforward action control scheme

It produces the following result:

$$\bar{u}_{\mathrm{ff}} = \frac{y_1}{K\tau}. \tag{6.3}$$

In this way, if the process is described perfectly by Model (6.1), an output transition in the time interval $[L, \tau + L]$ occurs. Then, at time $t = \tau + L$, the output settles at value y_1 thanks to the constant null value assumed by $u_{\mathrm{ff}}(t)$ for $t \geq \tau$.

Then, a suitable reference signal y_f has to be applied to the closed-loop system. It is desired that y_f be equal to the desired process output that would be obtained in the case where the process is modelled perfectly by Expression (6.1). Thus, the step reference signal r of amplitude y_1 has to pass through the system

$$F(s) = \frac{K\bar{u}_{\mathrm{ff}}}{y_1 s} e^{-Ls} \tag{6.4}$$

and then saturated at the level y_1.

It is worth stressing at this point that this method exploits the fact that a process output transition is required instead of tracking a general reference signal. In the latter case this scheme cannot be employed. The presence of many set-point changes can be instead easily handled by the PID plus nonlinear feedforward control system. Indeed, in case a new value of the set-point is selected during a previously determined transient response, it is sufficient to determine the feedforward action for the new value and to add it to the one that is currently applied. Analogously, the reference signal determined for the latest set-point change has to be added to the one related to the previous one.

The overall control scheme design involves the selection of the transition time τ and of the PID parameters. The choice of a sensible value of τ can be made by the user by considering that decreasing the value of τ means that the value of $\bar{u}_{\mathrm{ff}}$ (and therefore of the overall manipulated variable) increases, and a too low value of τ might imply that the determined control variable cannot be applied due to the saturation of the actuator. Thus, alternatively, the operator might first select the value of $\bar{u}_{\mathrm{ff}}$ depending on the desired control effort (defined typically as a percentage of the maximum limit of the manipulated variable) and determine consequently the value of τ. In this way the potentiality of the actuator can be fully exploited, and, in the nominal case, a time-optimal process variable transition (subject to the actuator constraints) is achieved. In any case, the design parameter τ has a clear physical meaning, because it handles the trade-off between performance, robustness, and control activity [54, 76] and can therefore be exploited to satisfy the specific requirements of a given application.

An appropriate tuning of the PID controller should take into account the robustness issue, since the feedforward action is based on a simple IPDT model of the plant and the compensation of the (unavoidable) modelling errors is left to the feedback control law. To this respect, it is very useful to consider the analysis made in [135], where it is shown that the deviations due to the modelling errors between the desired and the actual output can be treated as the effect of a load disturbance $d = G_d u_{\mathrm{ff}}$ where

$$G_d(s) = \frac{P(s) - F(s)}{P(s)}\big(1 + P(s)C(s)\big). \tag{6.5}$$

Thus, by considering that the process output results to be the superposition of the effects of the feedforward action and of the load disturbances (*i.e.*, of true load disturbances and the "fictitious" one d due to the modelling errors) and by considering also that in the nominal case the set-point following performance is determined only by the feedforward action, it is sensible to tune the PID controller by taking into account its load disturbance rejection performance. In this context, even if a time-optimal process variable transition is pursued, from a practical point of view it is worth considering a value of $\bar{u}_{\mathrm{ff}}$ which is lower than the true physical limit of the actuator so that the PID controller can handle the modelling uncertainties effectively.

6.2.2 Illustrative Examples

As a first illustrative example, consider the process

$$P(s) = \frac{1}{s}e^{-0.5s} \tag{6.6}$$

(*i.e.*, $K = 1$ and $L = 0.5$), and consider that a process variable from $y_0 = 0$ to $y_1 = 1$ is required. A maximum actuator limit $\bar{u}_{\mathrm{ff}} = 2.5$ is selected, and therefore the process variable transition results to be $\tau = y_1/(K\bar{u}_{\mathrm{ff}}) = 0.4$. The result obtained by employing the control scheme of Figure 6.1 is shown in Figure 6.2. Obviously, as the nominal case in considered, the output of the PID controller is zero (*i.e.*, $u = u_{\mathrm{ff}}$), and $y = y_f$. It appears that the time-optimal process variable transition is obtained as expected.

As a second example, in order to address modelling uncertainties, consider the process

$$P(s) = \frac{1}{s(0.1s+1)}e^{-0.5s} \tag{6.7}$$

and the same control task as before with $\bar{u}_{\mathrm{ff}} = 2.5$. By applying a standard relay-feedback identification method (see Section 2.2.2.1), the resulting ultimate gain and ultimate period are $K_u = 2.40$ and $P_u = 2.40$, respectively, and therefore the following model parameters are obtained (see Expressions (2.21) and (2.22)): $K = 0.88$ and $L = 0.6$. A PI controller tuned according to the Ziegler–Nichols rules (see Table 2.2) have been selected, namely, $K_p = 1.44$ and $T_i = 1.92$. The results obtained are shown in Figure 6.3.

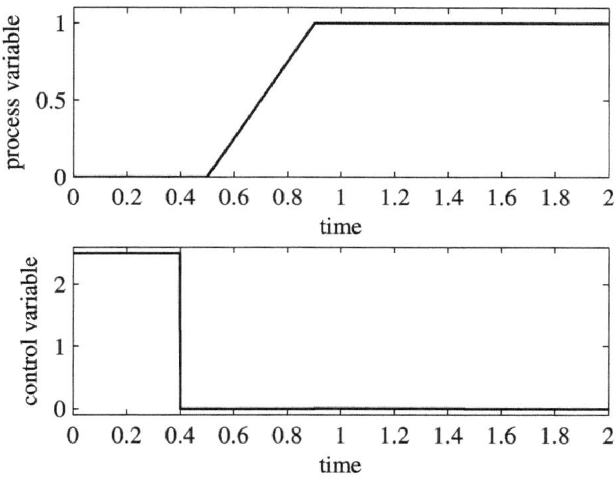

Fig. 6.2 Results obtained by applying the two-state time-optimal feedforward control law (nominal case)

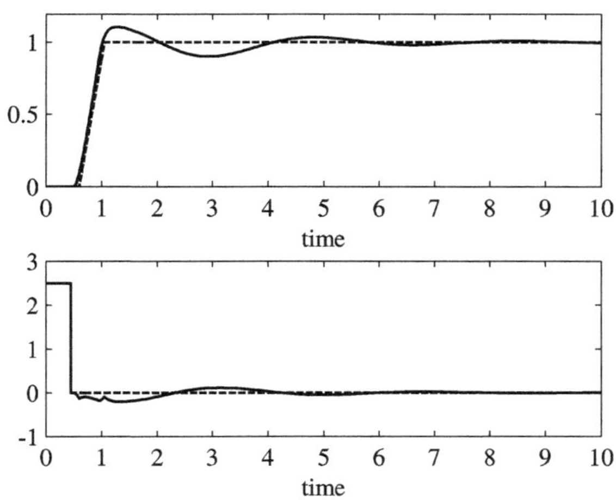

Fig. 6.3 Results obtained by applying the two-state time-optimal feedforward control law in the presence of modelling uncertainties. *Solid line*: process variable y (*top figure*) and control variable u (*bottom figure*). *Dashed line*: reference signal y_f (*top figure*) and feedforward signal u_{ff} (*bottom figure*)

6.3 Noncausal Feedforward Action: Continuous-time Case

The continuous-time case is first addressed in the context of noncausal feedforward control actions.

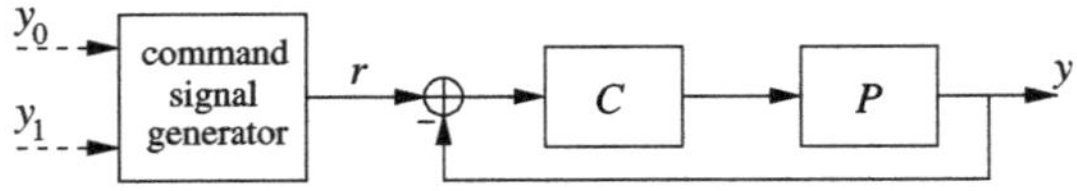

Fig. 6.4 Control scheme based on input–output inversion

6.3.1 Generalities

In the previous sections, it has been highlighted that the set-point following performance of a feedback control system can be significantly improved by the application of a properly designed (causal) feedforward action. From a different point of view, when the desired output trajectory is known in advance, a feedforward action determined by means of a stable inversion technique can be applied [167]. Roughly speaking, the approach consists in selecting a desired output function that meets the control requirements and then determining, by inverting the system dynamics, the input function that causes that selected output signal. It is worth noting that the concept of dynamic input–output inversion [20, 42] has been already proved to be effective in different areas of the automatic control field, such as motion control [92, 94], flight control [41], and robust control [95, 97].

In the context of PID control, the input–output inversion technique can be exploited to determine analytically a suitable command signal to be applied to the closed-loop control system, instead of the typical step signal, in order to achieve a high performance (*i.e.*, low rise time and low overshoot at the same time) when the process output is required to assume a new value. Indeed, assume that the process variable is required to achieve a steady-state value y_1 starting from a steady-state value y_0. As already mentioned, if a causal feedforward action is adopted, the control scheme of Figure 2.3 is based on the causal filtering of a step signal (of amplitude $y_1 - y_0$) by means of the system described by the transfer function $F(s)$. The resulting signal is then applied to the closed-loop system. Conversely, if an inversion approach is exploited, the scheme shown in Figure 6.4 is employed. In this case a step signal is not employed, but the knowledge in advance of y_1 is adopted by a command signal generator block to calculate a suitable command signal r to be applied to the closed-loop PID control system.

6.3.2 Modelling

The design methodology based on input–output inversion proposed in [99] is based on a theoretical framework that might appear to be somewhat complicated. However, the theoretical development can be made transparent to the user, and therefore the use of the technique does not impair the ease of use that is an essential requirement in the context of PID control. The fundamental passages are described hereafter in some detail in order to understand better the underlying concepts of the overall methodology.

As a first step of the devised method, the process to be controlled is modelled as an IPDT transfer function, *i.e.*,

$$P(s;\, K, L) = \frac{K}{s} e^{-Ls}, \tag{6.8}$$

but then, in order to have a rational transfer function, the dead-time term is approximated by means of a second-order Padè approximation. In this way, the approximated process transfer function results to be

$$\tilde{P}(s;\, K, T, L) = \frac{K}{s} \frac{1 - Ls/2 + L^2 s^2/12}{1 + Ls/2 + L^2 s^2/12}. \tag{6.9}$$

6.3.3 PID Controller Design

An output filtered PID controller in ideal form (2.4) is employed as a feedback controller. For the sake of clarity, its transfer function is recalled here:

$$C(s;\, K_p, T_i, T_d, T_f) = K_p \left(1 + \frac{1}{T s} + T_d s\right) \frac{1}{T_f s + 1}. \tag{6.10}$$

The tuning of the parameters can be done according to any of the many methods proposed in the literature (see Chapter 2) or even by a trial-and-error procedure. However, as the purpose of the overall procedure is the attainment of a high performance in the set-point following task, disregarding of the controller gains, it is sensible to select the PID parameters aiming only at obtaining a good load rejection performance.

6.3.4 Output Function Design

At this point, a desired output function that defines the transition from a set-point value y_0 to another y_1 (to be performed in the time interval $[0, \tau]$) has to be selected. Without loss of generality and for the sake of clarity, assume that $y_0 = 0$. A sensible choice is to adopt a so-called "transition" polynomial [96], *i.e.*, a polynomial function that satisfies boundary conditions and that is parameterised by the transition time τ. It is formally defined as

$$y_d(t) = c_{2k+1} t^{2k+1} + c_{2k} t^{2k} + \cdots + c_1 t + c_0. \tag{6.11}$$

The polynomial coefficients can be uniquely found by solving the following linear system, in which boundary conditions at the endpoints of interval $[0, \tau]$ are imposed:

$$\begin{cases} y_d(0) = 0;\ y_d(\tau) = y_1; \\ y_d^{(1)}(0) = 0;\ y_d^{(1)}(\tau) = 0; \\ \vdots \\ y_d^{(k)}(0) = 0;\ y_d^{(k)}(\tau) = 0. \end{cases} \tag{6.12}$$

The results can be expressed in closed-form as follows ($t \in [0, \tau]$):

$$y_d(t; \tau) = y_1 \frac{(2k+1)!}{k!} \sum_{i=k+1}^{2k+1} \frac{(-1)^{i-k-1}}{i(i-k-1)!(2k+1-i)!} \left(\frac{t}{\tau}\right)^i. \tag{6.13}$$

Expression (6.13) represents a monotonic function with neither undershooting nor overshooting, and its use is therefore very appealing in a practical context.

The order of the polynomial can be selected by imposing the order of continuity of the command input that results from the input–output inversion procedure [96]. Specifically, since the plant is modelled as an IPDT transfer function (see (6.8)), its relative degree is equal to one. Taking into account that the relative degree of the PID controller is zero, the relative degree of the overall closed-loop system is one. Thus, a third-order polynomial ($k = 1$) suffices if a continuous command input function is required, *i.e.*,

$$y_d(t; \tau) = y_1 \left(-\frac{2}{\tau^3} t^3 + \frac{3}{\tau^2} t^2 \right), \quad t \in [0, \tau]. \tag{6.14}$$

Outside the interval $[0, \tau]$, the function $y(t; \tau)$ is equal to 0 for $t < 0$ and to y_1 for $t > \tau$.

6.3.5 Stable Input–Output Inversion Algorithm

Once the closed-loop system is designed and the desired output function is selected, the problem of finding the command signal $r(t; K, L, K_p, T_i, T_d, T_f, \tau)$ that provides the desired output function has to be solved. For the sake of clarity of notation, the dependence of the functions and of the resulting coefficients from the parameters K, L, K_p, T_i, T_d, T_f is omitted in the following analysis. The closed-loop transfer function be denoted as

$$H(s) := \frac{C(s)\tilde{P}(s)}{1 + C(s)\tilde{P}(s)} = K_1 \frac{b(s)}{a(s)}, \tag{6.15}$$

where $b(s)$ and $a(s)$ are monic polynomials. As $H(s)$ is non-minimum phase (because of the positive zeros introduced by the Padè approximation, the straightforward inversion of the dynamics, *i.e.*, the calculation of $Y_d(s)/H(s)$, where $Y_d(s)$

is the Laplace transform of $y_d(t)$, would produce an unbounded command input function, which cannot be obviously adopted in practice. In other words, a stable dynamic inversion procedure is necessary, that is, a bounded input function has to be found in order to produce the desired output [98].

The numerator of the transfer function (6.15) can be rewritten as follows:

$$b(s) = b_-(s)b_+(s),$$

where $b_-(s)$ and $b_+(s)$ denote the polynomials associated to the zeros with negative real part (*i.e.*, those of the PID controller) and positive real part (*i.e.*, those of the Padè approximation), respectively. From (6.9) it can be derived that

$$b_+(s) = \left(s - z_R^+\right)^2 + z_I^{+^2}, \tag{6.16}$$

where $Z_R^+ = 3/L$ and $Z_I^+ = \sqrt{3}/L$, correspond to the complex zeros $z_R^+ \pm jz_I^+ \in \mathbb{C}_+$. From (6.10) three cases can be distinguished (depending on the selected PID parameters):

$$b_-(s) = (s - z_1^-)(s - z_2^-), \tag{6.17}$$

$$b_-(s) = (s - z^-)^2, \tag{6.18}$$

$$b_-(s) = (s - z_R^-)^2 + z_I^{-^2}, \tag{6.19}$$

corresponding to real distinct zeros (6.17), real coincident zeros (6.18), and complex zeros (6.19), respectively. Now, consider the inverse system of (6.15) whose transfer function can be written as:

$$H(s)^{-1} = \gamma_0 + \gamma_1 s + H_0(s),$$

where γ_0 and γ_1 are suitable constants, and $H_0(s)$, a strictly proper rational function, represents the zero dynamics. This can be uniquely decomposed according to

$$H_0(s) = H_0^-(s) + H_0^+(s) = \frac{c(s)}{b_-(s)} + \frac{d(s)}{b_+(s)},$$

where $c(s) = c_1 s + c_0$ and $d(s) = d_1 s + d_0$ are first-order polynomials with coefficients depending on K, L, K_p T_i, T_d, and T_f. The modes associated to $b^-(s)$ and $b^+(s)$ are denoted by $m_i^-(t)$, $i = 1, 2$, and by $m_i^+(t)$, $i = 1, 2$, respectively. More specifically, the unstable zero modes are given by

$$m_1^+(t) = e^{z_R^+ t} \cos z_I^+ t, \qquad m_2^+(t) = e^{z_R^+ t} \sin z_I^+ t, \tag{6.20}$$

while the stable zero ones are given according to the cases (6.17), (6.18), and (6.19) by

$$m_1^-(t) = e^{z_1^- t}, \qquad m_2^-(t) = e^{z_2^- t}, \tag{6.21}$$

$$m_1^-(t) = e^{z^- t}, \qquad m_2^-(t) = te^{z^- t}, \tag{6.22}$$

$$m_1^-(t) = e^{z_R^- t}\cos z_I^- t, \qquad m_2^-(t) = e^{z_R^- t}\sin z_I^- t. \tag{6.23}$$

With the Laplace transform operator $\mathcal{L}$, define

$$\eta_0^-(t) := \mathcal{L}^{-1}\big[H_0^-(s)\big]$$

and

$$\eta_0^+(t) := \mathcal{L}^{-1}\big[H_0^+(s)\big].$$

The following propositions and the following theorem represent the solution to the stable dynamic inversion problem.

Proposition 6.1

$$\int_0^t \eta_0^+(t-v)y_d(v;\tau)\,dv = H_0^+(0)y_d(t;\tau) + \frac{1}{\tau^3}\big(p_1^+(\tau)m_1^+(t) + p_2^+(\tau)m_2^+(t)\big)$$
$$+ \frac{1}{\tau^3}T_0^+(t;\tau), \tag{6.24}$$

where

$$T_0^+(t,\tau) = \begin{cases} s_0^+(t) + s_1^+(t)\tau & \text{if } t \in [0,\tau], \\ q_1^+(\tau)m_1^+(t-\tau) + q_2^+(\tau)m_2^+(t-\tau) & \text{if } t > \tau; \end{cases} \tag{6.25}$$

$p_i^+(\tau), q_i^+(\tau), i = 1, 2,$ *are suitable* τ*-polynomials; and* $s_i^+(t), i = 0, 1,$ *are suitable* t*-polynomials.*

Proposition 6.2

$$\int_0^t \eta_0^-(t-v)y_d(v;\tau)\,dv = H_0^-(0)y_d(t;\tau) + \frac{1}{\tau^3}\big(p_1^-(\tau)m_1^-(t) + p_2^-(\tau)m_2^-(t)\big)$$
$$+ \frac{1}{\tau^3}T_0^-(t,\tau), \tag{6.26}$$

where

$$T_0^-(t,\tau) = \begin{cases} s_0^-(t) + s_1^-(t)\tau & \text{if } t \in [0,\tau], \\ q_1^-(\tau)m_1^-(t-\tau) + q_2^-(\tau)m_2^-(t-\tau) & \text{if } t > \tau; \end{cases} \tag{6.27}$$

$p_i^-(\tau), q_i^-(\tau), i = 1, 2,$ *are suitable* τ*-polynomials; and* $s_i^-(t), i = 0, 1,$ *are suitable* t*-polynomials.*

Theorem 6.1 *The function* $r(t;\tau)$ *defined as*

$$r(t;\tau) = -\frac{1}{\tau^3}\big(p_1^+(\tau)m_1^+(t) + p_2^+(\tau)m_2^+(t)$$

$$- q_1^+(\tau)m_1^+(t-\tau) - q_2^+(\tau)m_2^+(t-\tau)\big) \quad \textit{if } t < 0, \tag{6.28}$$

$$\begin{aligned} r(t;\tau) &= \gamma_1 \dot{y}_d(t;\tau) + \gamma_0 y_d(t;\tau) + H_0(0) y_d(t;\tau) \\ &+ \frac{1}{\tau^3}\big(s_0^+(t) + s_0^-(t) + s_1^+(t)\tau + s_1^-(t)\tau - q_1^+(\tau)m_1^+(t-\tau) \\ &- q_2^+(\tau)m_2^+(t-\tau) + p_1^-(\tau)m_1^-(t) + p_2^-(\tau)m_2^-(t)\big) \quad \textit{if } t \in [0,\tau], \end{aligned} \tag{6.29}$$

$$\begin{aligned} r(t;\tau) &= \gamma_0 + H_0(0) + \frac{1}{\tau^3}\big(p_1^-(\tau)m_1^-(t) + p_2^-(\tau)m_2^-(t) \\ &+ q_1^-(\tau)m_1^-(t-\tau) + q_2^-(\tau)m_2^-(t-\tau)\big) \quad \textit{if } t > \tau \end{aligned} \tag{6.30}$$

is bounded over $(-\infty, +\infty)$, *and* $r(t;\tau)$ *causes the desired output* $y_d(t;\tau)$.

Proofs of the above propositions and of the above theorem can be found in [98]. Summarising, the determined function $r(t; K, L, K_p, T_i, T_d, T_f, \tau)$ exactly solves the stable inversion problem for FOPDT processes (in which the dead-time term has been substituted by a Padè approximation) controlled by a PID controller (6.10) and for a family of output functions, which depend on the free transition time τ. Actually, from a practical point of view, since the synthesised function (6.28)–(6.30) is defined over the interval $(-\infty, +\infty)$, it is necessary to adopt a truncated function $r_a(t;\tau)$, resulting therefore in an approximate generation of the desired output $y_d(t;\tau)$.

In particular, a preactuation time t_s and a postactuation time t_f can be selected so that the truncated (and therefore approximated) input $r_a(t;\tau)$ is equal to zero for $t < t_s$ and equal to $y_1/H(0)$ for $t > t_f$. By taking into account that the preactuation and postactuation inputs (*i.e.*, the input defined for $t < 0$ and $t > \tau$, respectively) converge exponentially to zero as time $t \to -\infty$ and to y_1 as time $t \to +\infty$, an arbitrarily precise approximation can be accomplished [96]. In particular, t_s and t_p can be calculated with arbitrary precision by selecting two arbitrary small parameters ε_0 and ε_1 and by subsequently determining

$$t_0 := \max\big\{t' \in \mathbb{R} : \big|r(t)\big| \le \varepsilon_0 \ \forall t \in (-\infty, t']\big\} \tag{6.31}$$

and

$$t_1 := \min\left\{t' \in \mathbb{R} : \left|r(t) - \frac{1}{H(0)}\right| \le \varepsilon_1 \ \forall t \in [t', \infty)\right\}. \tag{6.32}$$

Then, it has to be fixed

$$t_s = \min\{0, t_0\}, \qquad t_f = \max\{\tau, t_1\}. \tag{6.33}$$

Alternatively, in order to simplify the computation, the method suggested in [92] can be adopted. It consists in selecting

$$t_s = -\frac{10}{D_{\mathrm{rhp}}}, \qquad t_f = \tau + \frac{10}{D_{\mathrm{lhp}}}, \tag{6.34}$$

where D_{rhp} and D_{lhp} are the minimum distances of the right and left half-plane zeros respectively from the imaginary axis of the complex plane. Note that by taking into account (6.16), where $Z_R^+ = 3/L$, it is $t_s = -10L/3$. Hence, the approximate command signal to be actually used is

$$r_a(t;\tau) := \begin{cases} 0 & \text{for } t < t_s, \\ r(t;\tau) & \text{for } t_s \le t \le t_f, \\ y_1/H(0) & \text{for } t > t_f. \end{cases}$$

It is worth highlighting that the preactuation time depends only on the (apparent) dead time of the process, as this determines the unstable zeros of the closed-loop systems by means of the Padè approximation. Conversely, the postactuation time depends on the tuning of the PID parameters because the stable zeros of the closed-loop systems are those of the controller.

6.3.6 Discussions

The presented stable input–output inversion procedure can be performed by means of a symbolic computation, *i.e.*, a closed-form expression of the command input function $r(t; K, L, K_p, T_i, T_d, T_f, \tau)$ results. Indeed, the actual command signal to be applied for a given plant and a given controller is determined by substituting the actual value of the parameters into the resulting closed-form expression, and this actually motivates its strong appeal in the context of PID control. In this framework, the choice of using a second-order Padè approximation is motivated, from one side, by keeping the expression of $r(t; K, L, K_p, T_i, T_d, T_f, \tau)$ as simple as possible and, from the other side, by providing an approximation as good as possible, since the basic rationale of this method is to apply a model-based feedforward control action.

In any case, it is worth noting that the presented inversion procedure is based on a general one [98], where $H(s)$ can be the rational transfer function of any (stable) system, provided that there are not purely imaginary zeros. The method can be therefore trivially extended to PI, P, and PD control. In addition, the proposed approach can also be applied to (integral) processes with a high-order transfer function, as it is based on the inversion of the dynamics of the closed-loop system $H(s)$. However, in this case, the inversion procedure has to be performed on purpose. Conversely, if an IPDT model is employed, the determined general closed-form expression of $r(t; K, L, K_p, T_i, T_d, T_f, \tau)$ can be used.

Once the PID controller has been tuned, the only free design parameter is the transition time τ. Its role is basically the same as the transition time in the causal nonlinear feedforward method described in Section 6.2. That is, it handles the trade-off between performance, robustness, and control activity. It can be selected therefore by applying an analogous reasoning. However, since a closed-form expression of the control variable can be easily derived, the transition time can be also determined by solving an optimisation problem where its value has to be minimised subject to actuator constraints.

6.3.7 Practical Implementation

The command input in the interval $[t_s, t_f]$ is actually composed of three terms (the first one defined in the time interval $[t_s, 0]$, the second one in $[0, \tau]$, and the final one in $[\tau, t_f]$), and its expression, which depends on the time variable t, can be difficult to implement with standard industrial hardware/software components. For a simpler implementation of the method, it is convenient to obtain the command input as a step response, according to the standard two-degree-of-freedom control scheme of Figure 2.3. Therein, the signal to be applied to the closed-loop system is obtained as a step response of a filter $F(s)$. For this purpose, it is necessary first of all to shift the time axis by substituting $t = \bar{t} - t_s$ and by taking $\bar{t}$ as the new time variable. Then, the expression of the filter $F(s)$ can be obtained by applying the Laplace transform operator to $r_a(\bar{t}; \tau)$ [11]:

$$R_a(s; \tau) = \mathcal{L}\big[r_a(\bar{t}; \tau)\big] \tag{6.35}$$

and by imposing that

$$R_a(s; \tau) = \frac{1}{s} F(s; \tau). \tag{6.36}$$

Thus, it can be simply obtained that

$$F(s; \tau) = s R_a(s; \tau). \tag{6.37}$$

By performing the required (symbolic) computations and by substituting backwards $\bar{t} = t + t_s$, the command signal $r_a(t; \tau)$ is obtained as the step response of the following filters, to be considered in different time intervals:

$$F(s; \tau) = \begin{cases} \frac{\beta_{2,1}s^2+\beta_{1,1}s+\beta_{0,1}}{\alpha_{2,1}s^2+\alpha_{1,1}s+\alpha_{0,1}} & \text{for } 0 \le t < -t_s, \\ \frac{\beta_{7,2}s^7+\cdots+\beta_{0,2}}{\alpha_{7,2}s^7+\cdots+\alpha_{3,2}s^3} & \text{for } -t_s \le t < -t_s + \tau, \\ \frac{\beta_{2,3}s^2+\beta_{1,3}s+\beta_{0,3}}{\alpha_{2,3}s^2+\alpha_{1,3}s+\alpha_{0,3}} & \text{for } -t_s + \tau \le t < -t_s + t_f, \\ H^{-1}(0) & \text{for } t \ge -t_s + t_f, \end{cases} \tag{6.38}$$

where the polynomial coefficients $\beta_{2,1}, \ldots, \beta_{0,1}, \beta_{7,2}, \ldots, \beta_{0,2}, \beta_{2,3}, \ldots, \beta_{0,3}$ and $\alpha_{2,1}, \ldots, \alpha_{0,1}, \alpha_{7,2}, \ldots, \alpha_{3,2}, \alpha_{2,3}, \ldots, \alpha_{0,3}$ are resulting coefficients, whose analytical expression is not reported for brevity. In other words, a step signal has to be applied at $t = 0$ to the four different filters in (6.38), and then the command input to be applied to the closed-loop system is obtained by selecting the step responses of the filters according to the time intervals determined in (6.38). This strategy can be implemented easily in a Distributed Control System where a selector determines the required command input to be applied to the PID-based feedback control system by choosing between three transfer function blocks and a gain block according to the current time interval after the application of a set-point step signal. Note that there is no problem if the filter transfer functions are unstable because they are applied in a limited control interval.

A possible simplification for the implementation of the noncausal feedforward control strategy would be the use of a single transfer function instead of the four ones defined in (6.38). In general, the determined command input function can have a complex (non-monotonic) shape [99], and it can be difficult to represent it as a step response of a single transfer function. However, there are cases for which this can be possible with a good accuracy. In particular, this happens when the command input function is a smooth function. Conditions for the occurrence of this situation can be found by considering the following proposition [96].

Proposition 6.3 *Given the system* (6.15) *and the input function* (6.28)–(6.30), *the following limit holds*:

$$\lim_{\tau\to+\infty} \left\| H(0)r(\cdot;\tau) - y_d(\cdot;\tau) \right\|_{\infty} = 0, \tag{6.39}$$

where $\|f(\cdot)\| := \sup_{t\in\mathbb{R}} |f(t)|$ *denotes the* L_∞ *norm of a real signal* $f(t)$.

From a practical point of view this means that, when the transition time increases, the input function tends to be more similar to the desired output function. From another point of view, increasing the transition time τ yields a more robust system, namely, the obtained system output tends to be more similar to the desired output function. From these considerations it can be concluded that when the obtained system output has virtually no overshoot, then the corresponding system input is sufficiently smooth to be approximated as a step response of a single filter.

To confirm this fact and to derive a condition for the transition time, processes with different normalised dead time L/T controlled by PID controllers with the parameters tuned by different tuning rules have been considered. By considering a large number of simulation results with different systems and different controllers, it turns out that an overshoot of less than 5% is obtained in general when the transition time is selected as $\tau > 2L$. Thus, in this case the command input can be obtained as a step response of a single filter whose transfer function can be obtained by applying a standard least squares procedure [117] by taking a step signal as input and the determined command signal as output (note that there is no noise). In this context, it is worth determining the preactuation time interval as $t_s = -L$ in order to obtain a filter transfer function with no dead time. Further, many simulations have confirmed that a fourth-order transfer function is generally sufficient to obtain a satisfactory result.

6.3.8 Simulation Results

Consider the process

$$P(s) = \frac{0.0506}{s} e^{-6s} \tag{6.40}$$

and consider that the process output has to perform a transition from 0 to $y_1 = 1$. The PID parameters are selected, based on the minimisation of the integrated square error for set-point responses [128] (which is actually a PD controller with $K_p = 3.394$ and $T_d = 2.94$). For comparison, the filter time constant is selected as $T_f = 0.01$. The classic set-point step response is plotted in Figure 6.5. The noncausal feedforward action is determined by setting $\tau = 2L = 12$. The resulting value of the preactuation and postactuation times are $t_s = -20.01$ and $t_f = 41.4$. The determined command function is shown in Figure 6.6, and the corresponding control system responses are plotted in Figure 6.7. It appears that the process variable is virtually zero during the preactuation time, as expected, and then the transition occurs in a time interval of duration τ. In order to obtain the command input as a step response of a sequence of filters, the expression of $F(s)$ obtained by Laplace transforming directly $r_a(t)$ is

$$F(s) = \begin{cases} \frac{10^{-4}(0.05847s^2-0.13916s)}{s^2-s+0.33333} & \text{for } 0 \le t < -t_s, \\ \frac{p_2(s)}{q_2(s)} & \text{for } -t_s \le t < -t_s+\tau, \\ \frac{s^2+0.29245s+0.002215}{0.01994s^2+0.01329s+0.002215} & \text{for } -t_s+\tau \le t < -t_s+t_f, \\ 1 & \text{for } t \ge -t_s+t_f, \end{cases} \tag{6.41}$$

where

$$p_2(s) = 0.52232s^6 - 0.27729s^5 - 0.03015s^4 + 0.03446s^3 + 0.01801s^2 + 0.00019s - 0.00079 \tag{6.42}$$

and

$$q_2(s) = s^6 - 0.65986s^5 - 0.00680s^4 + 0.11338s^3. \tag{6.43}$$

The corresponding step response is virtually the same as the one shown in Figure 6.7. If the least squares procedure is applied in this case, the preactuation time is $t_s = -6$, and the resulting filter expression is

$$F(s) = \frac{0.05665s^4 - 0.01917s^3 + 0.08856s^2 + 0.00337s + 0.01318}{s^4 + 0.55118s^3 + 0.30330s^2 + 0.08502s + 0.01318}. \tag{6.44}$$

The corresponding step response is shown in Figure 6.8, and the resulting process and control variables are plotted in Figure 6.9. It can be seen that in this case the least squares procedures provides a similar result to the inversion-based technique.

Alternatively, if the tuning rules (2.134)–(2.136) with $\alpha = 1$ are considered, the following PID parameters are obtained: $K_p = 2.98$, $T_i = 15.66$, and $T_d = 1.93$ (the filter time constant is selected again as $T_f = 0.01$). The command signal obtained by applying the inversion procedure is shown in Figure 6.10 (note that $t_s = -20.01$ and $t_f = 146$), while the corresponding control system response is plotted in Figure 6.11 (for the set-point step response, see Figure 2.24). If the least squares approach is employed to approximate the command signal, the results are those shown

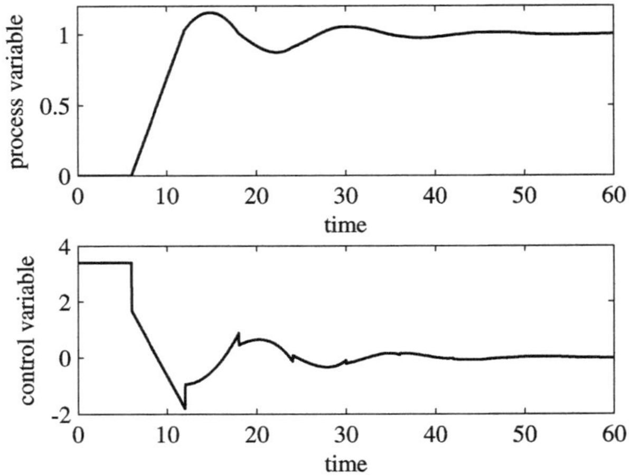

Fig. 6.5 Set-point step response with PD controller

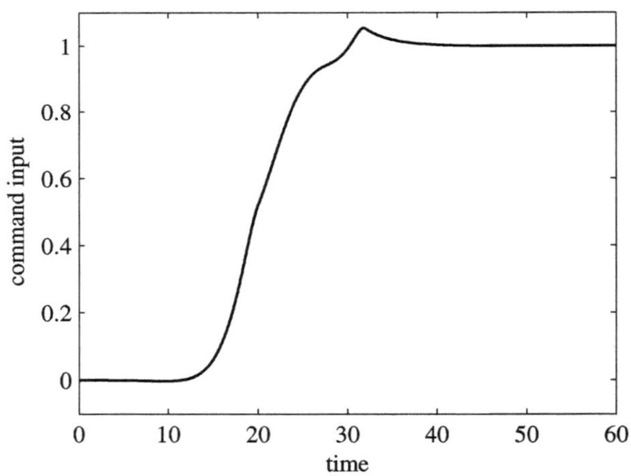

Fig. 6.6 Inversion-based command input with PD controller

in Figures 6.12 and 6.13. Note that a different preactuation time $t_s = -6$ has been employed in this latter case. It appears that the same considerations done for the PD controller case also apply to the PID controller case.

Finally, the role of the parameter τ is highlighted. For this purpose, consider the process (6.7) and the model obtained by applying the standard relay-feedback method (see Section 6.2.2) with $K = 0.88$ and $L = 0.6$. Then, as in Section 6.2.2, a PI controller tuned according to the Ziegler–Nichols rules (see Table 2.2) is selected, namely, $K_p = 1.44$ and $T_i = 1.92$. The command signals resulting by considering different values of the transition time $\tau = 0.6, 0.9, 1.2, 1.5$ are shown in Figure 6.14, while the corresponding process and control variables are shown in Figures 6.15

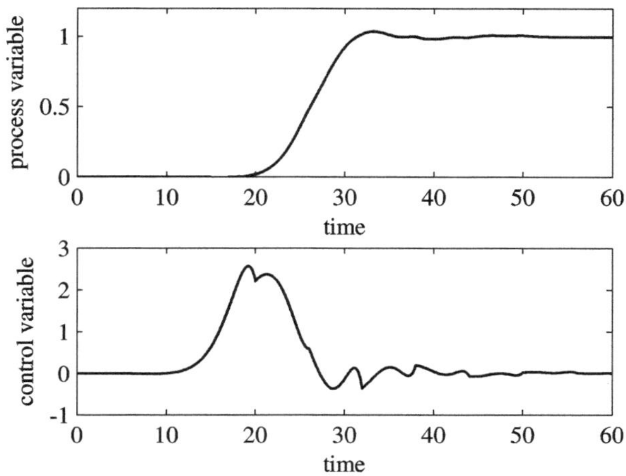

Fig. 6.7 Response with the noncausal feedforward approach with the PD controller

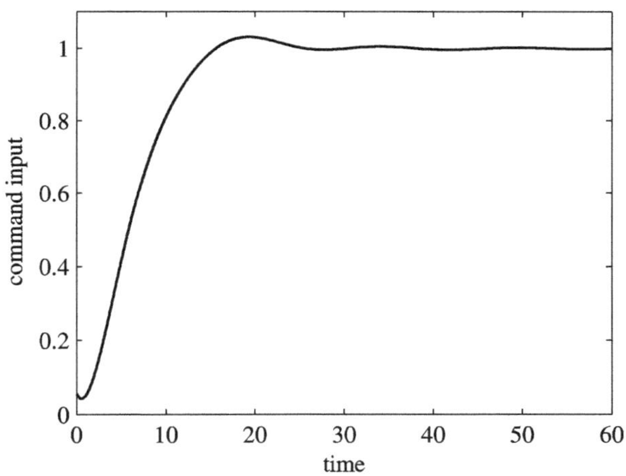

Fig. 6.8 Command input determined by means of a least squares procedure (PD controller)

and 6.16, respectively. It is evident that a sound choice of the transition time value handles the trade-off between aggressiveness and robustness (and control effort) well.

6.4 Noncausal Feedforward Action: Discrete-time Case

A noncausal feedforward control action can be devised also in the discrete-time case, as it is illustrated hereafter.

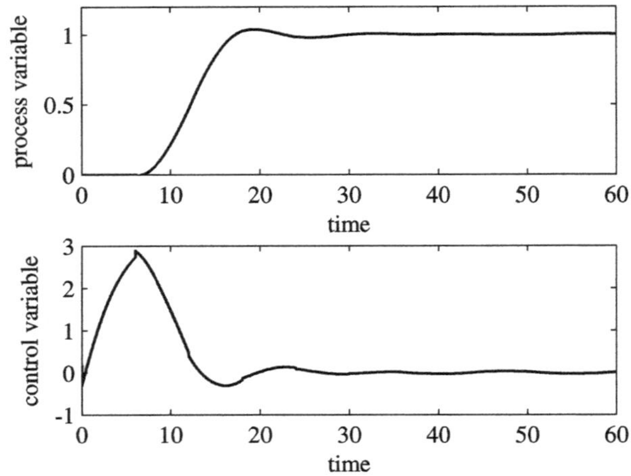

Fig. 6.9 Response with the use of a set-point filter determined by means of a least squares procedure (PD controller)

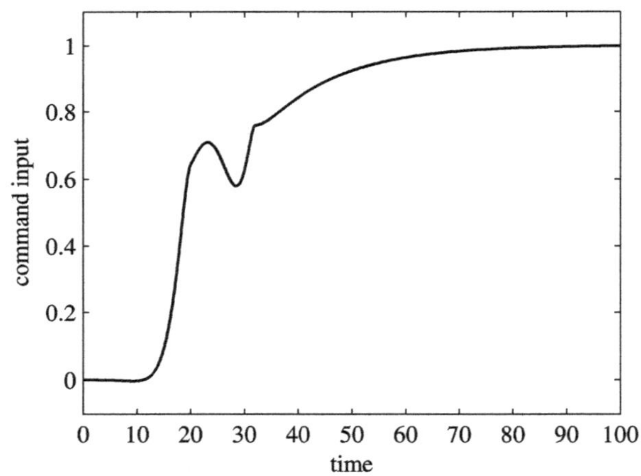

Fig. 6.10 Inversion-based command input with PID controller

6.4.1 Methodology

A noncausal feedforward action can be designed also in a different context, namely, by inverting the dynamics of the closed-loop system after having identified it in the discrete-time framework by means of a step response [134].

Consider again the scheme shown in Figure 6.4. Actually, the controller C can be of any type (provided that the closed-loop system is asymptotically stable), but for the sake of simplicity, it is assumed that it is a PID controller. As for the method described in Section 6.3, the aim is to find the command function $r(t)$ that produces a desired system output transition from y_0 to y_1, starting from time $t = 0$, but here

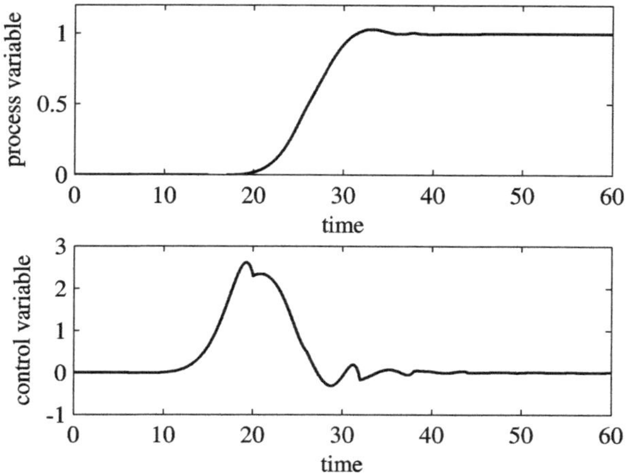

Fig. 6.11 Response with the noncausal feedforward approach with the PID controller

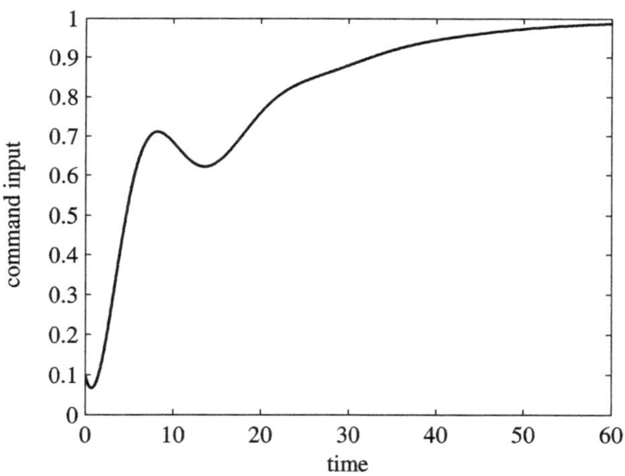

Fig. 6.12 Command input determined by means of a least squares procedure (PID controller)

no a priori knowledge on the process model is assumed. Despite the process and the controller being defined in the continuous-time domain, sampled data are considered in the following analysis (actually, nowadays the use of microprocessors is the common practice in industrial environments). It is assumed that the sampling time T has been chosen suitably by any standard technique [7].

An identification experiment can be easily performed by applying a step signal to the input of the closed-loop system. A closed-loop system model can then be

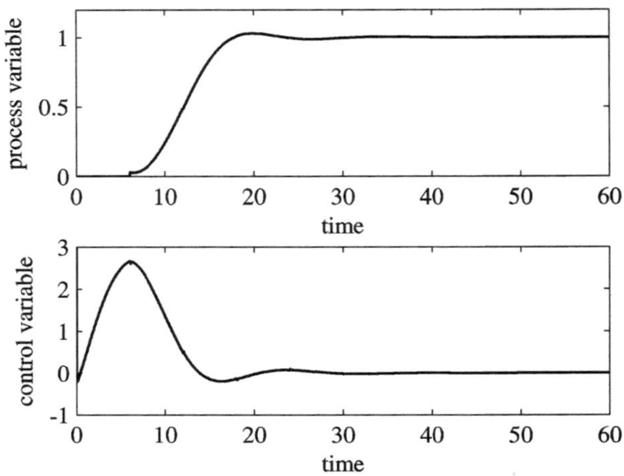

Fig. 6.13 Response with the use of a set-point filter determined by means of a least squares procedure (PID controller)

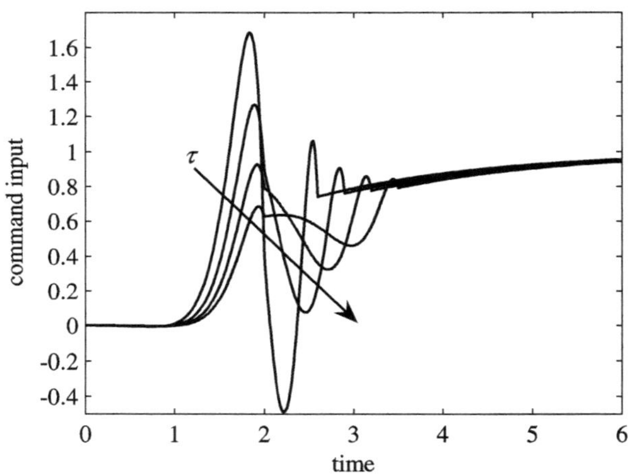

Fig. 6.14 Command signals for different values of the transition time τ

obtained by considering the truncated response ($t \in \{T, 2T, \ldots, NT\}$):

$$y(t) = y_0 + g_{t/T} r(0) + \sum_{i=1}^{\frac{t}{T}-1} g_i \big[r(t - iT) - r\big(t - (i+1)T\big)\big], \tag{6.45}$$

where $g_i := g(iT), i = 1, \ldots, N$, are the sampled output values in response to a unit-step input (see Figure 6.17), and $r(t)$ is the system input. For the sake of simplicity and without loss of generality, assume that $y_0 = 0$. The number N of parameters has to be taken sufficiently high in order to allow for a sufficiently accurate

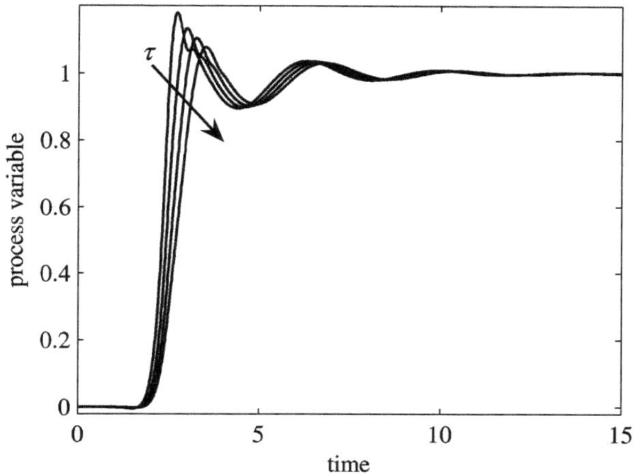

Fig. 6.15 Process variables for different values of the transition time τ

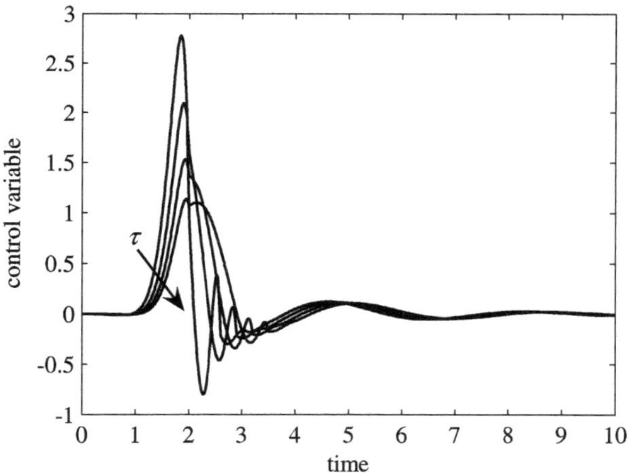

Fig. 6.16 Control variables for different values of the transition time τ

description of the system, but not too high to minimise the computational effort of the control strategy. From a practical point of view, the sampling of the step response in order to obtain parameters g_i should stop when the process output remains close to its steady-state value for a sufficiently long time.

For the presented methodology, it is convenient to write Expression (6.45) in the matrix form

$$Y = GR, \tag{6.46}$$

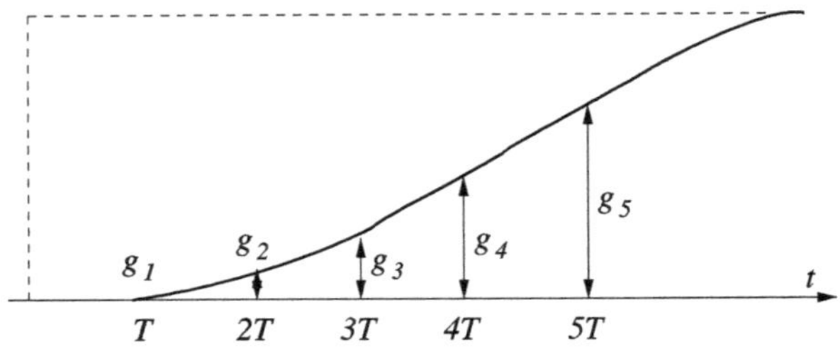

Fig. 6.17 Model coefficients based on step response

where

$$Y = \begin{bmatrix} y(T) \\ y(2T) \\ y(3T) \\ \vdots \\ y(NT) \end{bmatrix},$$

$$G = \begin{bmatrix} g_1 & 0 & 0 & \dots & 0 \\ -g_1 + g_2 & g_1 & 0 & \dots & 0 \\ -g_2 + g_3 & -g_1 + g_2 & g_1 & \dots & 0 \\ \vdots & \vdots & \vdots & \ddots & 0 \\ -g_{N-1} + g_N & -g_{N-2} + g_{N-1} & -g_{N-3} + g_{N-2} & \dots & g_1 \end{bmatrix},$$

and

$$R = \begin{bmatrix} r(0) \\ r(T) \\ r(2T) \\ \vdots \\ r((N-1)T) \end{bmatrix}.$$

It is worth noting that in many cases it might not be necessary to perform an ad hoc identification experiment (*i.e.*, to stop the normal process operations) in order to apply the presented methodology. In fact, as the model is obtained by evaluating a standard closed-loop step response, data taken from an output transition performed during routine process operations can be adopted. Obviously, it is important that the collected data be representative of a true step response (and therefore operations such as filtering and detrending might be necessary [56]) and if an unmeasured load disturbance occurs during the transient response, they should not be adopted. In this context, it can be useful to adopt the methods proposed in [32, 125] to detect load disturbances.

The desired output function is chosen again as a transition polynomial (6.13). In contrast with the continuous-time case, here its order can be chosen arbitrarily. Indeed, the order of the polynomial can be selected in order to handle the trade-off between the need to decrease the rise time and the need to decrease the control effort, taking into account that the rise time decreases and the control effort increases when the order of the polynomial increases. In general, a good choice in this context is to select $k = 2$, *i.e.*, the desired output function is

$$y_d(t;\tau) = \begin{cases} y_1(\frac{6}{\tau^5}t^5 - \frac{15}{\tau^4}t^4 + \frac{10}{\tau^3}t^3) & \text{if } 0 \le t \le \tau, \\ y_1 & \text{if } t > \tau. \end{cases} \tag{6.47}$$

Regarding the choice of the value of the transition time τ, the same considerations done in the continuous-time case can be applied also in this case.

Once the desired output function has been selected, *i.e.*, the array Y_d has been constructed, then the corresponding closed-loop system input $r(t)$ that causes $y_d(t;\tau)$ can be easily determined by simply inverting the system using Expression (6.46). In order for matrix G to be invertible by a standard numeric algorithm, it should be well conditioned, for example, there must not be a row (or a column) where all the elements are very small with respect to the elements of other rows (or columns). This happens when the process has a true dead time or an apparent dead time (*i.e.*, when the process is of high order), which causes some of the first sampled output values g_i of the step response to be null or almost null. Thus, denote by k the number of the first rows of G in which all the elements are less than a selected threshold ε. Then, matrix $\hat{G}$ can be obtained by removing the first k rows and the last k columns from G. Subsequently, by evaluating $y_d(t;\tau)$ at the first $N-k$ sampling time intervals, the array $Y_d = [y_d(T;\tau)\ y_d(2T;\tau)\ \cdots\ y_d((N-k)T;\tau)]^T$ can be easily constructed. The first $N-k$ values of the command reference input are then determined by applying the following expression:

$$\hat{R} = \left[r(T)\, r(2T)\ \cdots\ r\big((N-k)T\big)\right]^T = \hat{G}^{-1}Y_d. \tag{6.48}$$

In this way, the input function can be calculated by simply determining the inverse of a matrix, which can be performed by using different algorithms (see, for example, [103]).

Note that if the sampling time T and the value of N have been selected appropriately, as well as the value of τ, then the last element of the array $\hat{R}$ actually corresponds to the steady-state value of the input, and therefore the value of $r((N-k)T)$ can be applied to the closed-loop system for $t > (N-k)T$. Note also that, since the first k rows and the last k columns have been removed from matrix G, the output function obtained is delayed by kT with respect to the desired one. Actually, the dead time is removed in the model of the closed-loop system transfer function adopted in the dynamic inversion.

In the presence of measurement noise, as is always the case in practical applications, the method can be successfully applied, provided that the step response function employed for the identification of the closed-loop system model (6.45) is appropriately filtered. Since the required filtering can be performed off-line, a zero-phase noncausal filter can be applied in order to avoid a phase distortion. Further, the presence of the noise has to be considered when matrix $\hat{G}$ is constructed from G.

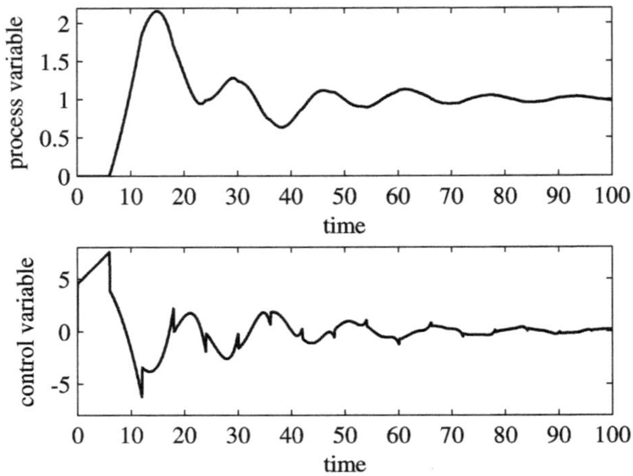

Fig. 6.18 Initial set-point step response employed for the discrete-time inversion approach

Actually, due to the noise measurements, it is sensible to redefine parameter ε as a noise band NB [8], *i.e.*, a threshold value that determines, as before, whether the sampled value g_i has to be discarded. Specifically, if $|g_i| < NB$, then g_i is considered to be zero in the construction of matrix $\hat{G}$. The value of NB can be easily selected by monitoring that process output for a sufficiently long time when the process is at steady-state.

Finally, it is worth stressing that, as in the continuous-time case, the inversion-based design of the feedforward action is independent of the PID design. It is therefore convenient to tune the controller in order to guarantee good load disturbance performance if this is of concern. In fact, even if this implies that the predicted closed-loop step response is unsatisfactory, the feedforward action is capable of providing an output transition with a low rise time and a low overshoot.

6.4.2 An Illustrative Example

As an illustrative example, consider again Process (6.40) with a PID controller tuned by minimising the ISE performance index for the load disturbance (see Table 2.5): $K_p = 4.51$, $T_i = 8.94$, and $T_d = 3.54$. The closed-loop step response is shown in Figure 6.18. By using a sampling interval $T = 0.02$ and a noise band $NB = 0.01$ and by setting a desired transition time $\tau = 20$, the discrete-time inversion technique procedure, applied to the closed-loop step response, gives the command input function shown in Figure 6.19. The resulting control system response is plotted in Figure 6.20. It appears that the performance has been improved significantly.

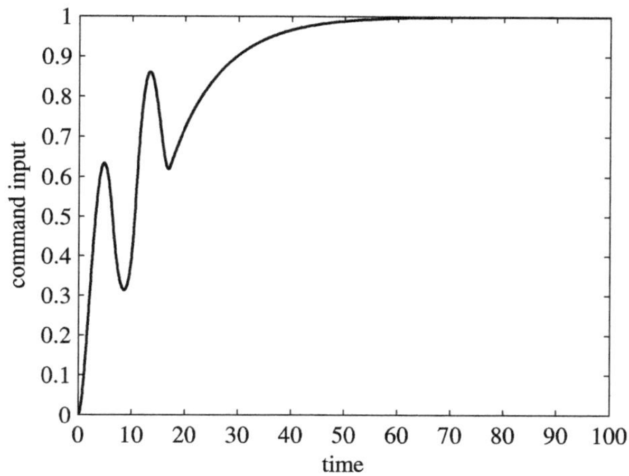

Fig. 6.19 Command input determined by the discrete-time inversion procedure

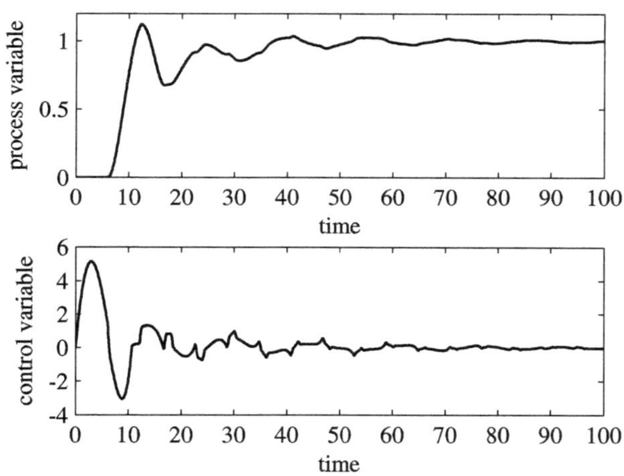

Fig. 6.20 Response obtained after having applied the discrete-time inversion procedure

6.5 Conclusions

A feedforward controller can be very beneficial in solving the problem of achieving a satisfactory performance both in the set-point following and in the load disturbance rejection task. Different methodologies for the design of a feedforward controller have been described in this chapter. In particular, the set-point following performance have been addressed. Features of the standard approach have been discussed. It has been shown that when the control task does not involve the tracking of a reference signal but only the transition of the process variable from a set-point value to another one is of concern, different alternative methods can be considered. The use of a nonlinear feedforward action improves the system performance considerably by taking into account explicitly the actuator constraints. Its implementation requires indeed a modest extra design effort. The great advantage of the noncausal approach is that a predefined performance can be actually obtained almost "independently" of the tuning of the (PID) controller and of the actual process dynamics. In fact, very similar responses are obtained with very different values of the PID parameters (namely, with PID parameters that provide very different set-point step responses) and with processes of different dynamics. This advantage is paid by the increased implementation complexity. It can be remarked that each of the considered methodologies has in any case a tuning parameter that handles the trade-off between aggressiveness and robustness.

Finally, it is worth highlighting that, if only the set-point following task is of concern, a fine tuning of the controller could lead to high performance, and the improvement provided by the use of a feedforward control system is not significant. However, the selection of the correct parameters can be difficult and time consuming. In this context the feedforwarded control action can be used to achieve (in a relatively easy way) satisfactory performance despite a not very appropriate tuning of the PID parameters and therefore to reduce the overall design effort. It is important to note that, in contrast with the constant set-point weighting approach, the use of a feedforward action can recover the set-point following performances even in the case of a sluggish tuning of the PID controller.

Chapter 7
PID–PD Control

An effective and simple way of designing a two-degree-of-freedom control scheme for an integral process is to employ an inner feedback loop in order mainly to stabilise the system and then an outer feedback loop in order to provide the required performance. Different methods have been proposed in the literature for the design and tuning of this control scheme, where the controllers are of P(D) type for the internal loop and of PI(D) type for the external loop. These methods will be presented and compared in this chapter.

7.1 The Control Scheme

The two degree-of-freedom control scheme considered in this chapter is the one shown in Figure 7.1 and consists in an internal loop, where the controller C_2 is mainly employed to stabilise the IPDT process P, and an external loop, where the controller C_1 is mainly devoted to satisfy the required performance. This control scheme can be easily rearranged in order to be the standard two-degree-of-freedom control scheme of Figure 2.3. The result is depicted in Figure 7.2, where it appears that, with reference to Figure 2.3, it is $C = C_1 + C_2$ and $F = C_1/(C_1 + C_2)$. Usually, C_2 is selected as a P or PD controller, while C_1 is selected as a PI or PID controller. Methodologies on how to tune the different controller parameters are described in the following sections.

7.2 PI–PD Structure

In this section, a few methods are considered where the inner loop controller has a PD structure (the derivative filter is omitted for the sake of simplicity), namely,

$$C_2(s) = K_{p2}(T_{d2}s + 1), \tag{7.1}$$

A. Visioli, Q.-C. Zhong, *Control of Integral Processes with Dead Time*,
Advances in Industrial Control,
DOI 10.1007/978-0-85729-070-0_7, © Springer-Verlag London Limited 2011

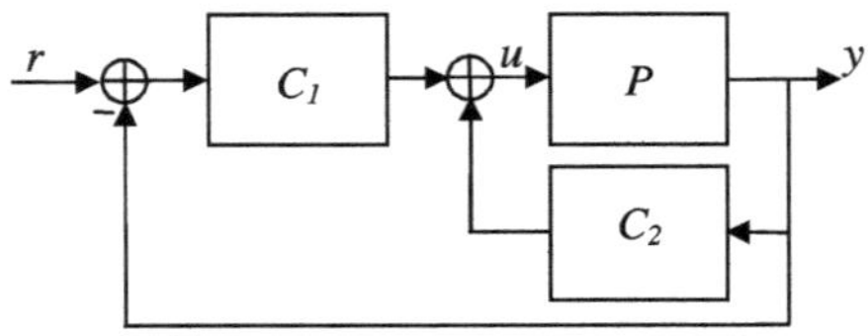

Fig. 7.1 Block diagram of the two-degree-of-freedom control scheme with a PID–PD structure

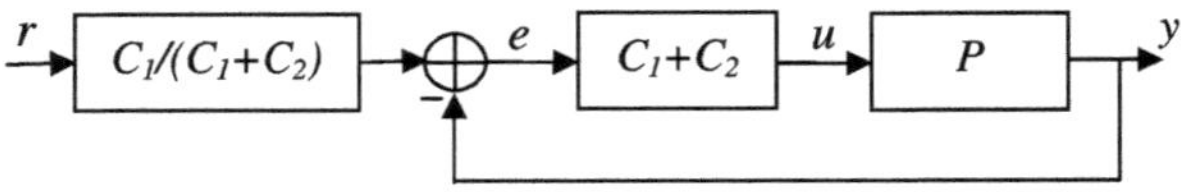

Fig. 7.2 Equivalent standard two-degree-of-freedom control scheme

and the outer loop controller has a PI structure, namely,

$$C_1(s) = K_{p1}\left(1 + \frac{1}{T_{i1}s}\right). \tag{7.2}$$

Methods for the tuning of the parameters are presented hereafter.

7.2.1 A Simple Approach

The simplest approach for the design of a PI–PD control structure is converting an existing standard one-degree-of-freedom PID controller [50]. In particular, consider Figure 7.2 and suppose that the feedback controller is of PID type, namely,

$$C(s) = K_p\left(1 + \frac{1}{T_i s} + T_d s\right). \tag{7.3}$$

By considering that $C(s) = C_1(s) + C_2(s)$ it can be derived that

$$C(s) = (K_{p2} + K_{p1})\left(1 + \frac{K_{p1}}{K_{p1} + K_{p2}}\frac{1}{T_{i1}s} + \frac{K_{p2}}{K_{p1} + K_{p2}}T_{d2}s\right). \tag{7.4}$$

In order to obtain the four PI–PD parameters from the three PID parameters, it is necessary to supply an additional relation between the parameters. This can be selected as $K_{p1} = \alpha K_{p2}$, so that, by comparing Expressions (7.3) and (7.4), they result to be

$$K_{p1} = \frac{\alpha}{1+\alpha}K_p, \tag{7.5}$$

$$K_{p2} = \frac{1}{1+\alpha}K_p, \tag{7.6}$$

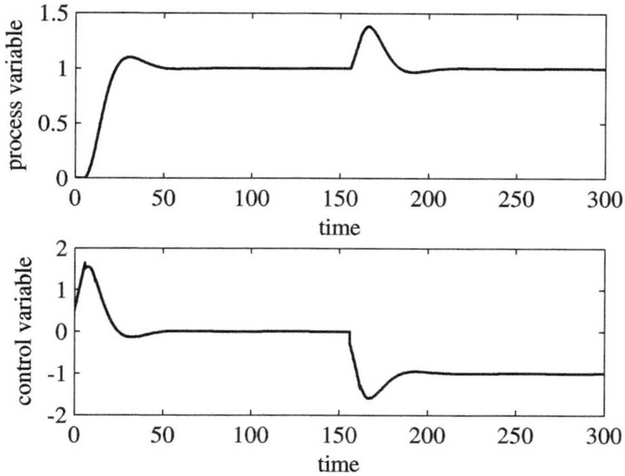

Fig. 7.3 Results obtained with the simple method based on the conversion of a standard PID controller

$$T_{i1} = \frac{\alpha}{1+\alpha} T_i, \tag{7.7}$$

$$T_{d2} = (1+\alpha) T_d. \tag{7.8}$$

It appears that the determined parameters heavily depend on the value of α. In [50] it is suggested, after having considered a large number of simulations, to choose $\alpha = 0.2$.

The set-point filter assumes the following form:

$$\begin{aligned} F(s) &= \frac{K_{p1}(T_{i1}s+1)}{K_{p2}T_{i1}T_{d2}s^2 + (K_{p1}T_{i1} + K_{p2}T_{i1})s + K_{p1}} \\ &= \frac{bT_i s + 1 + \alpha}{(T_i T_d + T_i T_d \alpha)s^2 + (\alpha T_i + T_i)s + 1 + \alpha}. \end{aligned} \tag{7.9}$$

As an illustrative example, consider the process

$$P(s) = \frac{0.0506}{s} e^{-6s}, \tag{7.10}$$

and consider the initial tuning obtained by applying the method described in Section 2.3.3.3, which yields $K_p = 2.98$, $T_i = 15.66$, and $T_d = 1.93$. Thus, the following PI–PD parameters are obtained: $K_{p1} = 0.497$, $T_{i1} = 2.610$, $K_{p2} = 2.483$, and $T_{d2} = 2.316$. The set-point and load disturbance step response is shown in Figure 7.3. By comparing the result with the one shown in Figure 2.24 it appears that, as expected, the use of the second degree-of-freedom reduces the overshoot significantly.

7.2.2 Tuning Method Based on the Standard Forms

In [47, 49] it has been proposed to determine the PI–PD controller parameters by minimising an integral performance criterion. In particular, the ISTE criterion defined as

$$J = \int_0^\infty te(t)\,dt, \tag{7.11}$$

where $e(t)$ is the control error, can be considered. In this context, it is possible to employ the so-called ISTE standard form, namely, a closed-loop transfer function which, assuming a process transfer function with no zeros (this issue will be discussed below) and a controller with one zero, can be written as

$$\bar{H}(s) = \frac{c_1 s + 1}{s^3 + d_2 s^2 + d_1 s + 1}. \tag{7.12}$$

The transfer function parameters d_1 and d_2 are selected in order to minimise the integral criterion (7.11) for different values of c_1. In this context, it can be shown that as c_1 increases, the performance index J decreases. However, the improvement in the step response becomes negligible for values of c_1 greater than four.

In order to employ the standard form concept for the tuning of the controller parameters, it is first necessary to approximate the IPDT process transfer function as

$$P(s) = \frac{K}{s} e^{-Ls} \cong \frac{K}{s(1+Ls)}. \tag{7.13}$$

Then, the closed-loop transfer function is determined as

$$H(s) = \frac{KK_{p1}(T_{i1}s+1)}{T_{i1}Ls^3 + (1+KK_{p2}T_{d2})T_{i1}s^2 + (K_{p1}+K_{p2})KT_i s + KK_{p1}}. \tag{7.14}$$

By comparing Expressions (7.12) and (7.14) it appears that a normalised standard form

$$\bar{H}(s_n) = \frac{c_1 s_n + 1}{s_n^3 + d_2 s_n^2 + d_1 s_n + 1} \tag{7.15}$$

can be obtained by fixing

$$s_n = \frac{s}{\alpha}, \tag{7.16}$$

$$\alpha = \left(\frac{KK_{p1}}{T_{i1}L}\right)^{\frac{1}{3}}, \tag{7.17}$$

$$c_1 = \alpha T_{i1}, \tag{7.18}$$

$$d_2 = \frac{1+KK_{p2}T_{d2}}{L\alpha}, \tag{7.19}$$

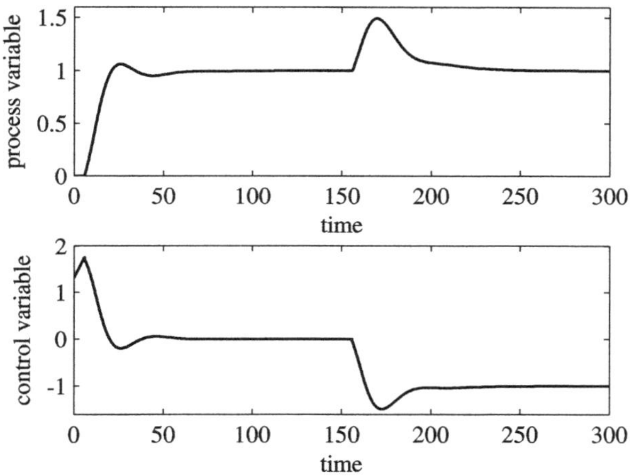

Fig. 7.4 Results obtained with the method based on standard forms

$$d_1 = \frac{(K_{p1} + K_{p2})K}{L\alpha^2}. \tag{7.20}$$

It is worth noting that the time scaling by α reduces the value of J by α^3. Equations (7.16)–(7.20) are therefore exploited, starting from appropriate values of K_{p1} and T_{i1}, in order to determine the value of α and the values of the PD controller parameters K_{p2} and T_{d2}.

As an illustrative example, consider again process (7.10). If the values of the PI parameters are selected as $K_{p1} = 1.31$ and $T_{i1} = 17.6$, from (7.17) the value $\alpha = 0.0856$ is determined, and then from (7.18) the value $c_1 = 1.507$ is obtained. By minimising J for this value of c_1, the values $d_1 = 2.637$ and $d_2 = 2.016$ are determined. Hence, from (7.19) and (7.20) it is easy to obtain $K_{p2} = 0.983$ and $T_{d2} = 0.695$. The set-point and load disturbance step response is shown in Figure 7.4.

It is worth stressing that the tuning technique can be employed, after an appropriate relay-feedback experiment, in the context of an automatic tuning methodology [47, 49].

7.3 PID–P Structure

Differently from Section 7.2, here a few methods are considered where the inner loop controller is just a proportional controller, namely,

$$C_2(s) = K_{p2}, \tag{7.21}$$

while the outer loop controller has a PID structure, which, for convenience, is expressed in parallel form as (the derivative filter is again not considered for the sake of simplicity)

$$C_1(s) = K_{p1} + \frac{K_{i1}}{s} + K_{d1}s. \tag{7.22}$$

It has to be remarked that in this case, the control variable can be written as (see Figure 7.1)

$$\begin{aligned} U(s) &= \left(K_{p1} + \frac{K_{i1}}{s} + K_{d1}s\right)\big(R(s) - Y(s)\big) - K_{p2}Y(s) \\ &= (K_{p1} + K_{p2})\left(\frac{K_{p1}}{K_{p1} + K_{p2}}R(s) - Y(s)\right) + \left(\frac{K_{i1}}{s} + K_{d1}s\right)E(s), \end{aligned} \tag{7.23}$$

which is the output of a parallel PID controller with set-point weight β, where

$$K_p = K_{p1} + K_{p2}, \tag{7.24}$$

$$K_i = K_{i1}, \tag{7.25}$$

$$K_d = K_{d1}, \tag{7.26}$$

and

$$\beta = \frac{K_{p1}}{K_{p1} + K_{p2}}. \tag{7.27}$$

Thus, in this case the control scheme is indeed a standard PID controller with set-point weight. The methods described in this section have been however situated in this part of the book because in their rationale the two-degree-of-freedom concept is exploited explicitly.

7.3.1 Tuning Method Based on Sensitivity Specifications

The design methodology proposed in [141] considers the SOIPDT process transfer function

$$P(s) = \frac{K}{s(Ts+1)}e^{-Ls}. \tag{7.28}$$

By employing a simple proportional controller $C_2(s) = K_{p2}$ in the inner loop, the inner closed-loop transfer function is

$$H_2(s) = \frac{Ke^{-Ls}}{Ts^2 + s + KK_{p2}e^{-Ls}}. \tag{7.29}$$

If the dead time term in the denominator is approximated, by using a Taylor series expansion, as

$$e^{-Ls} \cong 1 - Ls + 0.5L^2s^2, \tag{7.30}$$

then Expression (7.29) becomes

$$H_2(s) = \frac{Ke^{-Ls}}{(T + 0.5KK_{p2}L)s^2 + (1 - KK_{p2}L)s + KK_{p2}}. \tag{7.31}$$

In order for $H_2(s)$ to be stable, it has to be

$$K_{p2} < \frac{1}{KL}. \tag{7.32}$$

A suitable choice, which satisfies Expression (7.32) and provides optimum disturbance rejection [116], is

$$K_{p2} = \frac{0.2}{KL}. \tag{7.33}$$

With this choice, the inner closed-loop transfer function becomes

$$H_2(s) = \frac{e^{-Ls}}{\frac{T+0.1L}{K}s^2 + \frac{0.8}{K}s + \frac{0.2}{KL}}. \tag{7.34}$$

The PID controller is designed in order to mainly address the robustness issue. For this purpose, the maximum sensitivity is chosen as a robustness measure. It is defined as

$$M_s = \max_{0 \le \omega < \infty} \left| \frac{1}{1 + C_1(j\omega)H_2(j\omega)} \right|, \tag{7.35}$$

and it represents the inverse of the shortest distance from the Nyquist curve of the loop transfer function $C_1(s)H_2(s)$ from the critical point $(-1, 0)$. Note that the maximum sensitivity is related to the phase margin ϕ_m and to the gain margin A_m by the following inequalities:

$$A_m > \frac{M_s}{M_s - 1} \tag{7.36}$$

and

$$\phi_m > 2 \arcsin \frac{1}{2M_s}. \tag{7.37}$$

At this point it is worth rewriting the PID controller transfer function as

$$C_1(s) = k\left(\frac{As^2 + Bs + C}{s} \right), \tag{7.38}$$

where

$$A = \frac{K_{d1}}{k}, \qquad B = \frac{K_{p1}}{k}, \qquad C = \frac{K_{i1}}{k}. \tag{7.39}$$

Then, the controller zeros are chosen in order to cancel the poles of the process seen by the controller, namely, $H_2(s)$. This yields

$$A = \frac{T + 0.1L}{K}, \qquad B = \frac{0.8}{K}, \qquad C = \frac{0.2}{KL}, \tag{7.40}$$

and the open-loop transfer function becomes

$$G(s) = C_1(s)H_2(s) = \frac{k}{s}e^{-Ls}. \tag{7.41}$$

In order for $G(s)$ to be tangential to the circle centred in $(-1, 0)$ and with radius $1/M_s$ (*i.e.*, in order for the system to have a maximum sensitivity M_s), the following conditions must hold (see Figure 7.5):

$$G(j\omega) = -1 + \frac{1}{M_s}e^{-j\theta}, \tag{7.42}$$

$$\arg\frac{dG(j\omega)}{d\omega} = \frac{\pi}{2} - \theta, \tag{7.43}$$

that is, by considering (7.41),

$$\frac{k}{\omega}\sin(\omega L) = 1 - \frac{1}{M_s}\cos\theta, \tag{7.44}$$

$$\frac{k}{\omega}\cos(\omega L) = \frac{1}{M_s}\sin\theta, \tag{7.45}$$

$$\omega L + \arctan\frac{1}{\omega L} = \frac{\pi}{2} + \theta. \tag{7.46}$$

The nonlinear system of Equations (7.44)–(7.46) has no analytical solution. However, an approximate solution for parameter k can be determined numerically as [141]

$$k = \frac{1}{L}\left(1.451 - \frac{1.508}{M_s}\right). \tag{7.47}$$

The PID parameters are therefore given as

$$K_{p1} = \frac{1}{KL}\left(1.1608 - \frac{1.2064}{M_s}\right), \tag{7.48}$$

$$K_{i1} = \frac{1}{KL^2}\left(0.29 - \frac{0.3016}{M_s}\right), \tag{7.49}$$

$$K_{d1} = \frac{T + 0.1L}{KL}\left(1.450 - \frac{1.508}{M_s}\right). \tag{7.50}$$

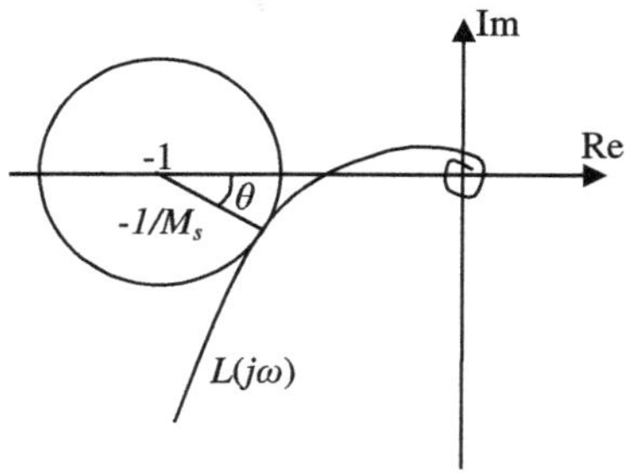

Fig. 7.5 Explanation of the maximum sensitivity concept

Summarising, Expressions (7.33) and (7.48)–(7.50) provide the tuning of the two-degree-of-freedom controller once the maximum sensitivity parameter has been specified by the user. Note that if a standard PID controller with set-point weight is considered, by applying Expressions (7.24)–(7.27) it can be easily deduced that

$$K_p = \frac{1}{KL}\left(1.3608 - \frac{1.2064}{M_s}\right), \tag{7.51}$$

$$K_i = K_{i1}, \tag{7.52}$$

$$K_d = K_{d1}, \tag{7.53}$$

and

$$\beta = \frac{1.1608M_s - 1.2064}{1.3608M_s - 1.2064}. \tag{7.54}$$

As an illustrative example, consider the process

$$P(s) = \frac{1}{s(s+1)}e^{-0.2s}. \tag{7.55}$$

By applying the tuning rule (7.33) $K_{p2} = 1$ results, independently on the selected value for the maximum sensitivity. Then, if the typical (low) value $M_s = 1.4$ is chosen [5], by applying the tuning rules (7.48)–(7.50), it results $K_{p1} = 1.495$, $K_{i1} = 1.864$, and $K_{d1} = 1.901$ (if tuning rules (7.51)–(7.54) are employed, it is $K_p = 2.495$, $K_{i1} = 1.864$, and $K_{d1} = 1.901$, $\beta = 0.6$). The set-point and load disturbance step response is shown in Figure 7.6.

Alternatively, if the typical (high) value $M_s = 2.0$ is chosen [5], by applying the tuning rules (7.48)–(7.50), it results $K_{p1} = 2.788$, $K_{i1} = 3.480$, and $K_{d1} = 3.550$ (if tuning rules (7.51)–(7.54) are employed, it is $K_p = 3.788$, $K_{i1} = 3.480$, and $K_{d1} = 3.550$, $\beta = 0.736$). The corresponding result is plotted in Figure 7.7. It appears that, as expected, for a higher value of the maximum sensitivity, the control system is more aggressive, and therefore a better load disturbance rejection performance is paid by a higher control effort.

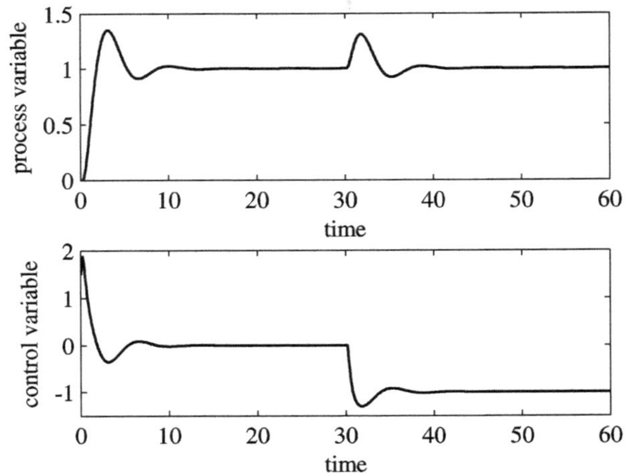

Fig. 7.6 Results obtained with the method based on sensitivity specifications ($M_s = 1.4$)

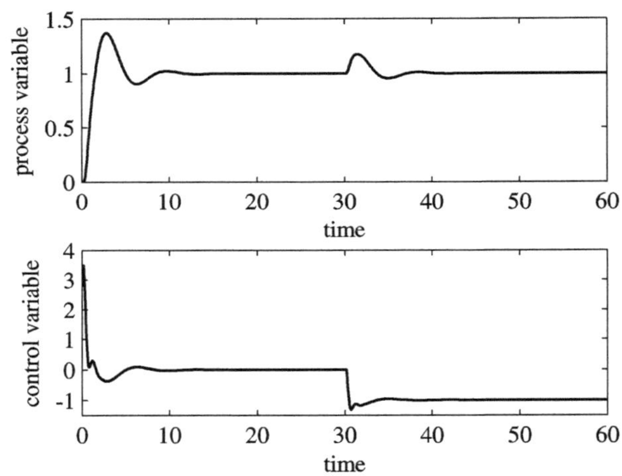

Fig. 7.7 Results obtained with the method based on sensitivity specifications ($M_s = 2.0$)

7.3.2 Tuning Method Based on Phase and Gain Margins

A possible alternative to the method described in Section 7.3.1 is to specify the required gain and phase margins instead of the required maximum sensitivity [142, 143]. This means that the same procedure of Section 7.3.1 is followed, but Equations (7.42) and (7.43) are substituted by four equations that impose that the loop transfer function (7.41) provides the desired gain margin A_m and phase margin ϕ_m. They are

$$\arg G(j\omega_p) = -\pi, \tag{7.56}$$

$$A_m \left| G(j\omega_p) \right| = 1, \tag{7.57}$$

$$\left| G(j\omega_g) \right| = 1, \tag{7.58}$$

$$\phi_m = \pi + \arg G(j\omega_g), \tag{7.59}$$

where w_g is the gain crossover frequency, and w_p is the phase crossover frequency. By considering Expression (7.41), this set of equations can be rewritten as

$$\omega_p L = \frac{\pi}{2}, \tag{7.60}$$

$$A_m = \frac{\omega_p}{k}, \tag{7.61}$$

$$k = \omega_g, \tag{7.62}$$

$$\phi_m = \frac{\pi}{2} - \omega_g L. \tag{7.63}$$

From Equations (7.61)–(7.62) it can be deduced that

$$A_m \omega_g = \omega_p. \tag{7.64}$$

Multiplying both sides of (7.64) by L and considering (7.60) and (7.63), the relation between the gain margin and the phase margin can be derived (note that this is the same result obtained in Section 3.1.3):

$$\phi_m = \frac{\pi}{2}\left(1 - \frac{1}{A_m}\right). \tag{7.65}$$

Then, by choosing the reasonable values $A_m = 3$ and $\phi_m = 60°$, from (7.60) and (7.61) the value of k can be obtained as

$$k = \frac{\pi}{2 A_m L} = \frac{\pi}{6L}. \tag{7.66}$$

Hence, the PID parameters can be determined as

$$K_{p1} = \frac{2\pi}{15KL}, \tag{7.67}$$

$$K_{i1} = \frac{\pi}{30KL^2}, \tag{7.68}$$

$$K_{d1} = \frac{\pi(T + 0.1L)}{6KL}. \tag{7.69}$$

The equivalent standard single-loop PID controller with set-point weight can be determined easily by considering Expressions (7.51)–(7.54) as

$$K_p = \frac{2\pi + 3}{15KL}, \tag{7.70}$$

$$K_i = K_{i1}, \tag{7.71}$$

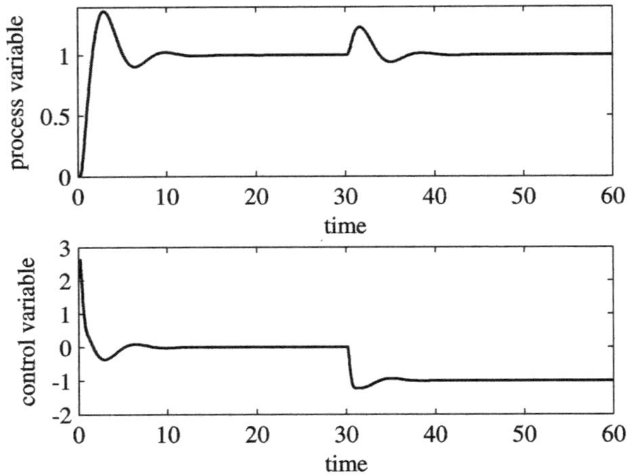

Fig. 7.8 Results obtained with the method based on phase and gain margins specifications

$$K_d = K_{d1}, \tag{7.72}$$

and

$$\beta = \frac{2\pi}{2\pi + 3}. \tag{7.73}$$

As an illustrative example, consider again process (7.55). By applying the tuning rules (7.33) and (7.67)–(7.69), the following parameters are obtained: $K_{p2} = 1$, $K_{p1} = 2.094$, $K_{i1} = 2.618$, $K_{d1} = 2.670$ (or, equivalently, $K_p = 3.094$, $K_i = 2.618$, $K_d = 2.670$, $\beta = 0.677$). The resulting set-point and load disturbance step response is shown in Figure 7.8.

7.3.3 Tuning Method Based on a New Robustness Specification

Alternatively to the maximum sensitivity and to the gain and phase margins, a new robustness measure λ, to be employed in the controller design, has been proposed in [149]. It is defined as the inverse of the maximum absolute real part of the open-loop transfer function (see Figure 7.9), namely,

$$\frac{1}{\lambda} = \max_{0 \le \omega < \infty} \left| \mathrm{Re}\big[G(j\omega)\big] \right|. \tag{7.74}$$

This new specification is related to the gain and phase margins by means of the following inequalities:

$$A_m > \lambda, \tag{7.75}$$

$$\phi_m > \arccos\left(\frac{1}{\lambda}\right). \tag{7.76}$$

Then, if the internal loop is considered with a process model (7.28) and a proportional controller $C_2(s) = K_{p2}$, the robustness index can be determined as

$$\begin{aligned}
\frac{1}{\lambda} &= \max_{0\le\omega<\infty}\left|\mathrm{Re}\left[\frac{KK_{p2}e^{-j\omega L}}{j\omega(j\omega T+1)}\right]\right| \\
&= \max_{0\le\omega<\infty}\left|-KK_{p2}\frac{\omega T\cos(\omega L)+\sin(\omega L)}{\omega(\omega^2T^2+1)}\right| \\
&= \lim_{\omega\to 0}\left|-KK_{p2}\frac{\omega T\cos(\omega L)+\sin(\omega L)}{\omega(\omega^2T^2+1)}\right| \\
&= KK_{p2}(T+L).
\end{aligned} \tag{7.77}$$

By choosing the reasonable value $\lambda = 2$ it is evidently

$$K_{p2} = \frac{0.5}{K(T+L)}. \tag{7.78}$$

With this choice, the inner closed-loop transfer function becomes

$$T_2(s) = \frac{e^{-Ls}}{\frac{T^2+TL+0.25L^2}{K(T+L)}s^2 + \frac{T+0.5L}{K(T+L)}s + \frac{0.5}{K(T+L)}}. \tag{7.79}$$

The same concept can be employed for the tuning of the controller C_1. By exploiting again a pole-zero cancellation as in Sections 7.3.1 and 7.3.2, the parameters of the PID controller (7.38) are determined as

$$A = \frac{T^2+TL+0.25L^2}{K(T+L)}, \tag{7.80}$$

$$B = \frac{T+0.5L}{K(T+L)}, \tag{7.81}$$

$$C = \frac{0.5}{K(T+L)}. \tag{7.82}$$

This yields the open-loop transfer function (7.41). In order to determine the parameter k, the same robustness measure as before can be used. Thus, it is

$$\begin{aligned}
\frac{1}{\lambda} &= \max_{0\le\omega<\infty}\left|\mathrm{Re}\left[\frac{ke^{-j\omega L}}{j\omega}\right]\right| \\
&= \max_{0\le\omega<\infty}\left|\frac{-k\sin(\omega L)}{\omega}\right| \\
&= \lim_{\omega\to 0}\left|\frac{-k\sin(\omega L)}{\omega}\right|
\end{aligned}$$

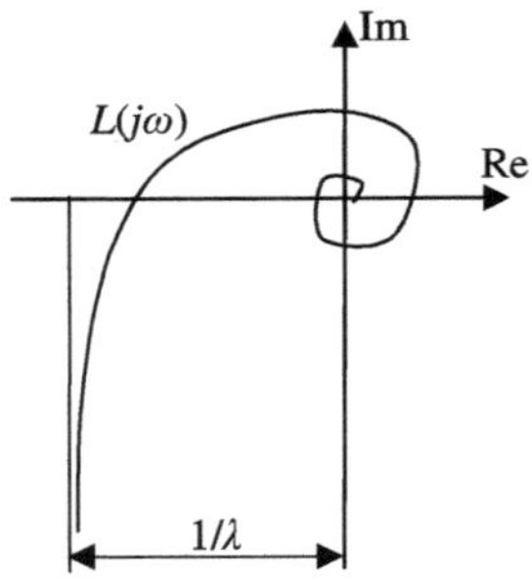

Fig. 7.9 Explanation of the λ robustness measure

$$\begin{aligned} &= \lim_{\omega\to 0}\left|\frac{-kL\cos(\omega L)}{1}\right| \\ &= kL. \end{aligned} \tag{7.83}$$

By choosing again $\lambda = 2$ it is evidently $k = 0.5/L$, and therefore the PID parameters are determined as (see (7.39) and (7.80)–(7.82))

$$K_{p1} = \frac{0.5(T + 0.5L)}{KL(T + L)}, \tag{7.84}$$

$$K_{i1} = \frac{0.25}{KL(T + L)}, \tag{7.85}$$

$$K_{d1} = \frac{0.5(T^2 + TL + 0.25L^2)}{KL(T + L)}. \tag{7.86}$$

Equivalently, the parameters of a standard PID controller with set-point weight are

$$K_p = \frac{0.5(T + 1.5L)}{KL(T + L)}, \tag{7.87}$$

$$K_i = \frac{0.25}{KL(T + L)}, \tag{7.88}$$

$$K_d = \frac{0.5(T^2 + TL + 0.25L^2)}{KL(T + L)}, \tag{7.89}$$

$$\beta = \frac{T + 0.5L}{T + 1.5L}. \tag{7.90}$$

The same process (7.55) is considered as an illustrative example. By applying the devised tuning rules, the following controller parameters can be determined: $K_{p2} = 0.420$, $K_{p1} = 2.292$, $K_{i1} = 1.042$, $K_{d1} = 2.521$ (or, equivalently, $K_p = 2.708$, $K_i = 1.042$, $K_d = 2.521$, $\beta = 0.677$). The resulting set-point and load disturbance step response is shown in Figure 7.10.

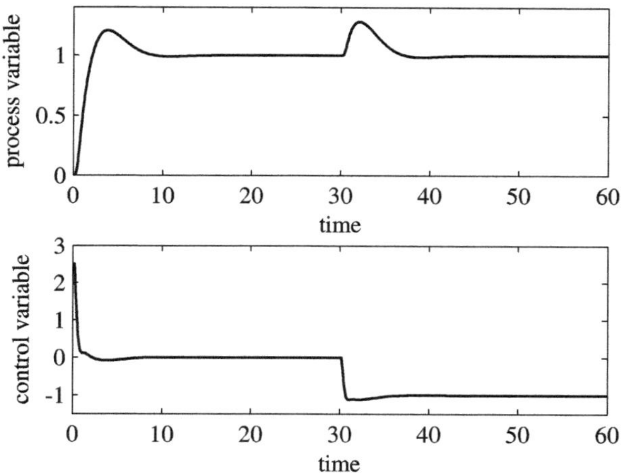

Fig. 7.10 Results obtained with the method based on a new robustness specification for a PID–P controller

7.4 PID–PD Structure

The last case addressed in this chapter is the use of a PD controller in the inner loop and a PID controller in the outer loop, namely (for convenience, the derivative filter is neglected),

$$C_2(s) = K_{p2}(1 + T_{d2}s), \tag{7.91}$$

$$C_1(s) = K_{p1} + \frac{K_{i1}}{s} + K_{d1}s. \tag{7.92}$$

In this case, the control variable can be written as (see Figure 7.1)

$$\begin{aligned} U(s) &= \left(K_{p1} + \frac{K_{i1}}{s} + K_{d1}s\right)\big(R(s) - Y(s)\big) - K_{p2}(1 + T_{d2}s)Y(s) \\ &= (K_{p1} + K_{p2})\left(\frac{K_{p1}}{K_{p1} + K_{p2}}R(s) - Y(s)\right) + \frac{K_{i1}}{s}E(s) \\ &\quad + (K_{d1} + K_{p2}T_{d2})s\left(\frac{K_{d1}}{K_{d1} + K_{p2}T_{d2}}R(s) - Y(s)\right), \end{aligned} \tag{7.93}$$

which is the output of a parallel PID controller with set-point weight β for the proportional action and a set-point weight γ for the derivative action (see Section 2.1.2), where

$$K_p = K_{p1} + K_{p2}, \tag{7.94}$$

$$K_i = K_{i1}, \tag{7.95}$$

$$K_d = K_{d1} + K_{d2}, \tag{7.96}$$

$$\beta = \frac{K_{p1}}{K_{p1} + K_{p2}}, \tag{7.97}$$

and

$$\gamma = \frac{K_{d1}}{K_{d1} + K_{p2}T_{d2}}. \tag{7.98}$$

7.4.1 Tuning Method Based on a New Robustness Specification

The method described in Section 7.3.3 can be extended to the case of PID–PD controller [150]. By considering the process model (7.28), the inner loop controller can be designed by applying a pole-zero cancellation, namely,

$$T_{d2} = T. \tag{7.99}$$

The inner loop transfer function is therefore

$$H_2(s) = C_2(s)P(s) = \frac{KK_{p2}}{s}e^{-Ls}, \tag{7.100}$$

and the controller gain K_{p2} can be determined based on the robustness specification λ (see Section 7.3.3). Indeed, it is

$$\begin{aligned} \frac{1}{\lambda} &= \max_{0\le\omega<\infty}\left|\mathrm{Re}\left[\frac{KK_{p2}e^{-j\omega L}}{j\omega}\right]\right| \\ &= \max_{0\le\omega<\infty}\left|\frac{-KK_{p2}\sin(\omega L)}{\omega}\right| \\ &= \lim_{\omega\to 0}\left|\frac{-KK_{p2}\sin(\omega L)}{\omega}\right| \\ &= \lim_{\omega\to 0}\left|\frac{-KK_{p2}L\cos(\omega L)}{1}\right| \\ &= KK_{p2}L. \end{aligned} \tag{7.101}$$

Hence, by selecting $\lambda = 2$, it is

$$K_{p2} = \frac{1}{2KL}. \tag{7.102}$$

With this PD controller, the inner closed-loop transfer function is

$$H_2(s) = \frac{Ke^{-Ls}}{Ts^2 + s + \frac{1}{2L}(Ts+1)e^{-Ls}}, \tag{7.103}$$

which, by approximating the dead time term in the denominator as $e^{-Ls} \cong 1 - Ls$, can be rewritten, after a few trivial passages, as

$$H_2(s) = \frac{e^{-Ls}}{\frac{T}{2K}s^2 + \frac{L+T}{2KL}s + \frac{1}{2KL}}. \tag{7.104}$$

The parameters of the PID controller (7.38) are determined, by applying again a pole-zero cancellation as

$$A = \frac{T}{2K}, \tag{7.105}$$

$$B = \frac{L+T}{2KL}, \tag{7.106}$$

$$C = \frac{1}{2KL}, \tag{7.107}$$

and this yields the open-loop transfer function (7.41). The parameter k can be therefore determined as in (7.83) by considering the robustness measure λ, yielding $\frac{1}{\lambda} = kL$, and therefore, by selecting $\lambda = 2$, it is $k = 0.5/L$. Thus, the PID tuning rules are

$$K_{p1} = \frac{0.25(T+L)}{KL^2}, \tag{7.108}$$

$$K_{i1} = \frac{0.25}{KL^2}, \tag{7.109}$$

$$K_{d1} = \frac{0.25T}{KL}. \tag{7.110}$$

Equivalently, the parameters of a standard PID controller with set-point weights (for both the proportional and derivative action) are

$$K_p = \frac{0.25(3L+T)}{KL^2}, \tag{7.111}$$

$$K_i = \frac{0.25}{KL^2}, \tag{7.112}$$

$$K_d = \frac{0.75T}{KL}, \tag{7.113}$$

$$\beta = \frac{T+L}{T+3L}, \tag{7.114}$$

$$\gamma = \frac{1}{3}. \tag{7.115}$$

Note that the set-point weight applied to the derivative action is independent of the process parameters.

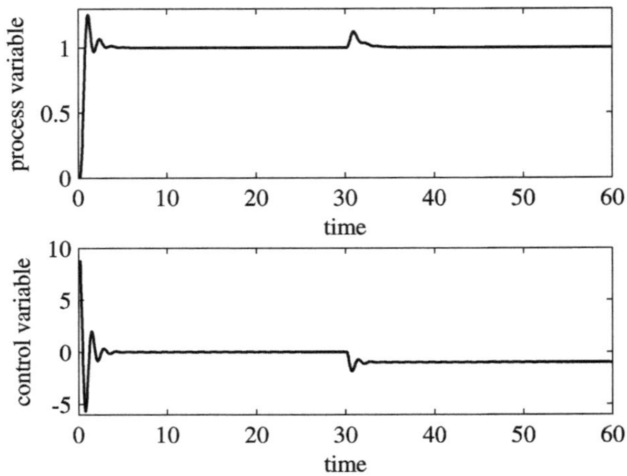

Fig. 7.11 Results obtained with the method based on a new robustness specification for a PID–PD controller

As an illustrative example, the same process (7.55) is considered. By applying the devised tuning rules, the following controller parameters can be determined: $K_{p2} = 2.5$, $T_{d2} = 1$, $K_{p1} = 7.5$, $K_{i1} = 6.25$, $K_{d1} = 1.25$ (or, equivalently, $K_p = 10$, $K_i = 6.25$, $K_d = 3.75$, $\beta = 0.75$, $\gamma = 1/3$). The resulting set-point and load disturbance step response is shown in Figure 7.11. By comparing this result with the one shown in Figure 7.10, it appears that the presence of the derivative action in the inner loop controller allows the design of a more aggressive controller in the outer loop and therefore to obtain a better load disturbance performance rejection without impairing significantly the set-point following performance. However, this is obtained at the expense of a more significant control effort.

7.4.2 A More Complex Controller

An alternative controller form, similar to the PID-PD structure, has been proposed in [66]. In this case, while a PD controller (7.91) is employed again for the inner loop, the controller of the outer loop has the following form (it will be shown hereafter that, actually, this is not a standard PID controller in series form):

$$C_1(s) = K_{p1}\left(1 + \frac{K_{i1}}{s}\right)(1 + T_{d1}s). \tag{7.116}$$

Regarding the controller C_2, by selecting again the derivative time constant of the filter equal to the time constant of the SOIPDT process, *i.e.*, $T_{d2} = T$, the open-loop transfer function (7.100) results. In this case, the proportional gain is selected by considering that the gain that stabilises the integral process can be derived, by applying the Nyquist stability criterion, as

$$K_{p2} = \frac{\frac{\pi}{2} - \phi_m}{KL}, \tag{7.117}$$

and therefore, an appropriate tuning is to select $\phi_m = 61.3°$, that is,

$$K_{p2} = \frac{0.5}{KL}. \tag{7.118}$$

Then, the internal closed-loop transfer function becomes

$$T_2(s) = \frac{Ke^{-Ls}}{(Ts+1)\left(s + \frac{e^{-Ls}}{2L}\right)}. \tag{7.119}$$

In order to perform a pole-zero cancellation, instead of approximating the dead time term, in [66] it is suggested to select

$$T_{d1} = T \tag{7.120}$$

and

$$K_{i1} = \frac{e^{-Ls}}{2L}. \tag{7.121}$$

Note that the presence of the delay in the integral action means that this controller is indeed different from a standard PID controller. In particular, it can be said that the integral action exhibits a dead-time compensation.

Finally, the proportional gain is tuned by addressing the robustness issue. Indeed, by considering that with the previous choice the overall loop transfer function is

$$W(s) = \frac{K_{p1}K}{s}e^{-Ls}, \tag{7.122}$$

relations (7.56)–(7.59) give the following expression:

$$K_{p2} = \frac{\phi_m}{(A_m - 1)KL}. \tag{7.123}$$

Although any values can be considered for the desired gain and phase margins, it is suggested to select $A_m = 3$ and $\phi_m = \pi/3$. This yields

$$K_{p2} = \frac{0.5236}{KL}. \tag{7.124}$$

The same process (7.55) is considered as an illustrative example. By applying the devised tuning rules, the following controller parameters can be determined: $K_{p2} = 2.5$, $T_{d2} = 1$, $K_{p1} = 2.618$, $K_{i1} = 2.5$, $T_{d1} = 1$. The resulting set-point and load disturbance step response is plotted in Figure 7.12. It appears that system achieves very good performance in both the set-point following and the load disturbance rejection.

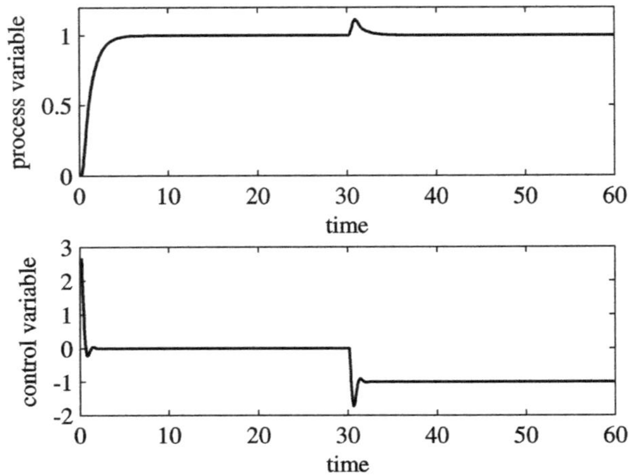

Fig. 7.12 Results obtained with the more complex controller

7.5 Conclusions

In this chapter it has been shown that the control of an IPDT process can be addressed by considering a two-degree-of-freedom control scheme where an internal loop with a P(D) controller is employed to mainly stabilise the system and then an external loop with a PI(D) controller is obtained to achieve the required performance. It has been highlighted that the overall control scheme can also be seen as a standard PID controller with set-point weight for the proportional action and, in case, for the derivative action. Thus, the implementation of the presented methodologies is not particularly difficult, also because the design techniques end up with tuning rules easy to apply. Finally, it has been shown that the presence of a dead time term in the integral action of the external controller yields a significant improvement in the performance (at the expense of an increased complexity of the controller).

Chapter 8
Smith-predictor-based Control

When the dead time of the integral process is significant, traditional control schemes such as those seen in the previous chapters might not be sufficient to obtain the required performance. In these cases, a dead time compensator can be considered. The most well-known control scheme where a dead time compensator is implemented is the Smith predictor, which, however, in its classical implementation, fails to provide a null steady-state error in the presence of a constant load disturbance if the process exhibits an integral dynamics. For this reason, mainly, different modifications of the classical Smith predictor have been proposed in the literature to overcome this drawback. These approaches will be reviewed and compared in this chapter.

8.1 Classical Smith Predictor

The presence of a significant dead time in a process prevents the achievement of a high-performance control system because it is difficult to obtain a high value of the gain crossover frequency (namely, a fast transient response) and a satisfactory phase margin (namely, a small overshoot) at the same time. In order to compensate the effect of the dead time, a dead time compensator can be designed. The Smith predictor is a one-degree-of-freedom control scheme that can be considered as the classical solution to be employed as a dead time compensator [114]. The control scheme is shown in Figure 8.1, where $\tilde{P}(s)$ is the delay-free part of the process. It appears that the presence of the model of the process is used in order to predict the future values of the output. In particular, if a perfect model of the process is employed, the feedback signal can be trivially expressed as $y(t+L)$ where L is the dead time of the process. In other words, the dead time is outside of the control loop, as it is shown in Figure 8.2, which represents, from the point of view of the set-point response, an equivalent scheme of Figure 8.1. In this way, the controller can be designed, in principle, as the process were delay free. Obviously, the unavoidable modelling uncertainties have to be taken into account, and therefore it is not advisable to design an aggressive controller as it could make the overall control system unstable.

A. Visioli, Q.-C. Zhong, *Control of Integral Processes with Dead Time*,
Advances in Industrial Control,
DOI 10.1007/978-0-85729-070-0_8, © Springer-Verlag London Limited 2011

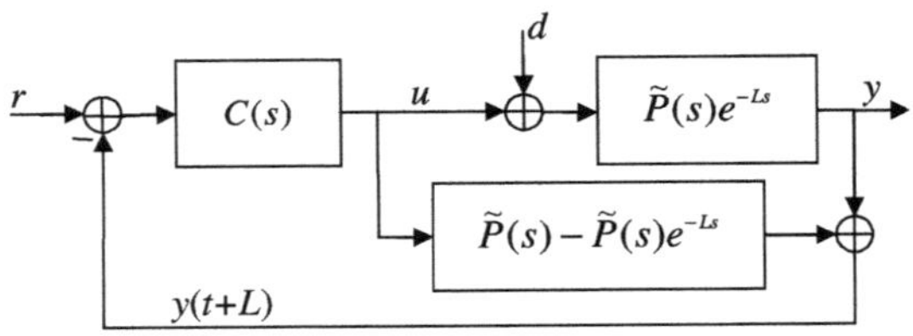

Fig. 8.1 Block diagram of the classical Smith predictor

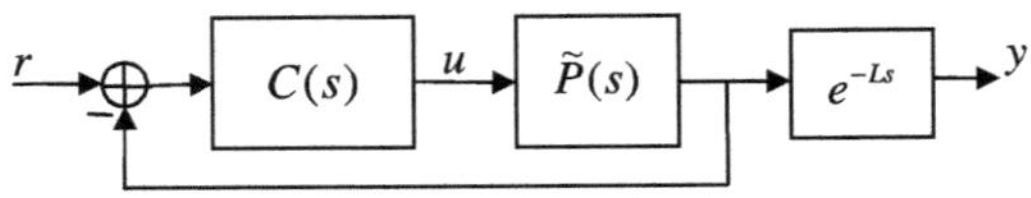

Fig. 8.2 Equivalent block diagram of the classical Smith predictor

In any case, while the classical Smith predictor is effective for self-regulating processes, when it is applied to integral processes, it is unable to provide a null steady-state error in the presence of a constant load disturbance d [16]. In fact, consider the IPDT process

$$P(s) = \frac{K}{s}e^{-Ls}, \tag{8.1}$$

for which, evidently, $\tilde{P}(s) = K/s$, and a PI controller

$$C(s) = K_p\left(1 + \frac{1}{T_i s}\right). \tag{8.2}$$

According to the control scheme of Figure 8.1, the process output can be calculated as

$$\begin{aligned} Y(s) = {} & \frac{K_p\big(1 + \frac{1}{T_i s}\big)\frac{K}{s}e^{-Ls}}{1 + K_p\big(1 + \frac{1}{T_i s}\big)\frac{K}{s}}R(s) \\ & + \frac{K}{s}e^{-Ls}\left(1 - \frac{K_p\big(1 + \frac{1}{T_i s}\big)\frac{K}{s}e^{-Ls}}{1 + K_p\big(1 + \frac{1}{T_i s}\big)\frac{K}{s}}\right)D(s), \end{aligned} \tag{8.3}$$

and therefore, by considering only the load disturbance response, the expression of the process output can be rewritten as

$$Y(s) = \frac{K(T_i s^2 + KK_pT_i(1 - e^{-Ls})s + KK_p(1 - e^{-Ls}))e^{-Ls}}{s(T_i s^2 + KK_pT_i s + KK_p)}D(s). \tag{8.4}$$

Hence, when a unit step signal is applied as a load disturbance (*i.e.*, $D(s) = 1/s$), by applying the final value theorem and the l'Hôpital's rule, it can be determined that the steady-state value of the process output is KL.

As an illustrative example, consider the process

$$P(s) = \frac{1}{s}e^{-5s} \tag{8.5}$$

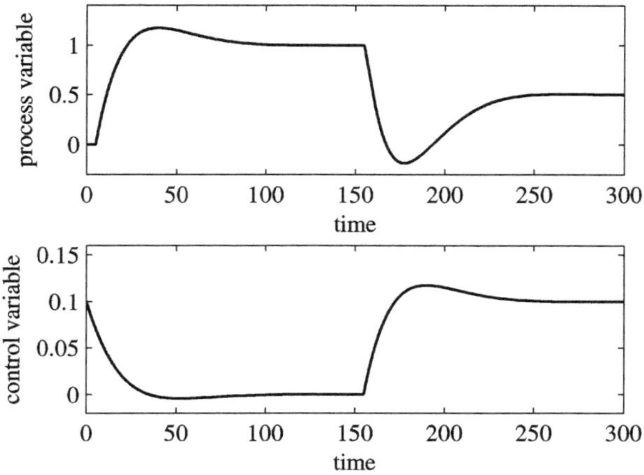

Fig. 8.3 Results obtained by applying the classical Smith predictor

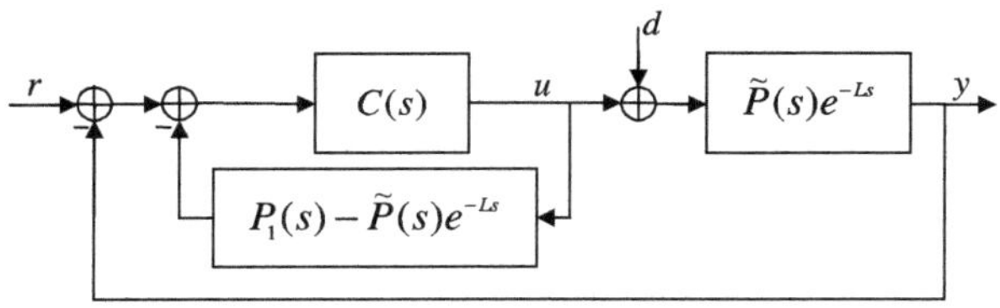

Fig. 8.4 Block diagram of the modified Smith predictor

and a PI controller with $K_p = 0.1$ and $T_i = 27$. The response of the classical Smith predictor of Figure 8.1 when a unit step is applied to the set-point at time $t = 0$ and a step of amplitude $d = -0.1$ is applied as a load disturbance at time $t = 150$ is shown in Figure 8.3, where it appears that a steady-state error occurs in the presence of a constant load disturbance.

In order to overcome this problem, different solutions have been proposed in the literature. They will be presented in the next sections.

8.2 Modified Smith Predictor

A modification of the original Smith predictor to cope with load disturbances has been proposed in [145]. It consists of modifying the scheme of Figure 8.1 and is shown in Figure 8.4 (note that it is still a one-degree-of-freedom controller).

In this case,

$$P_1(s) = K\frac{1 - Ls}{s}. \tag{8.6}$$

If $C(s)$ is a PI controller (see (8.2)), the transfer function between the set-point r and the process variable y is

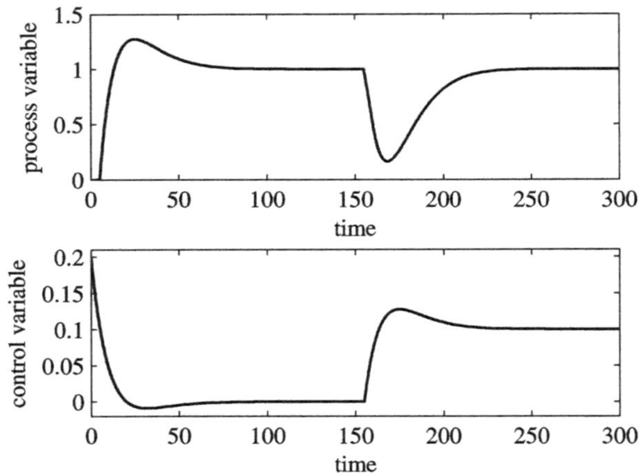

Fig. 8.5 Results obtained by applying the modified Smith predictor (nominal case)

$$
\begin{aligned}
H(s) = \frac{Y(s)}{R(s)} &= \frac{\frac{C(s)P(s)}{1+C(s)(P_1(s)-P(s))}}{1+\frac{C(s)P(s)}{1+C(s)(P_1(s)-P(s))}} \\
&= \frac{C(s)P(s)}{1+C(s)P_1(s)} \\
&= \frac{KK_p(T_i s+1)e^{-Ls}}{T_i(1-KK_pL)s^2+KK_p(T_i-L)s+KK_p}, \qquad (8.7)
\end{aligned}
$$

while the transfer function between the load disturbance d and the process variable y is

$$
\begin{aligned}
H_d(s) = \frac{Y(s)}{D(s)} &= \frac{P(s)}{1+\frac{C(s)P(s)}{1+C(s)(P_1(s)-P(s))}} \\
&= \frac{K(T_i(1-KK_pL)s^2+KK_p(T_i-L-T_ie^{-Ls})s+KK_p(1-e^{-Ls}))e^{-Ls}}{s(T_i(1-KK_pL)s^2+KK_p(T_i-L)s+KK_p)}.
\end{aligned}
\qquad (8.8)
$$

It can be easily determined (by applying the final value theorem) that, with this modification, the controller is capable to provide a null steady-state error in the presence of a constant load disturbance. For example, consider again process (8.5) and a PI controller (as in Section 8.1) with $K_p = 0.1$ and $T_i = 27$. The response of a unit step set-point signal and of a step load disturbance of amplitude $d = -0.1$ is plotted in Figure 8.5.

The null steady-state error is achieved also in the presence of modelling uncertainties. For example, when there is a 10% estimation error in the dead time (namely, the value of the dead time term employed in the controller is $L = 5.5$), the response is shown in Figure 8.6. The control scheme is effective, but since it is a one-degree-of-freedom control scheme, it is difficult to achieve high performance in both the

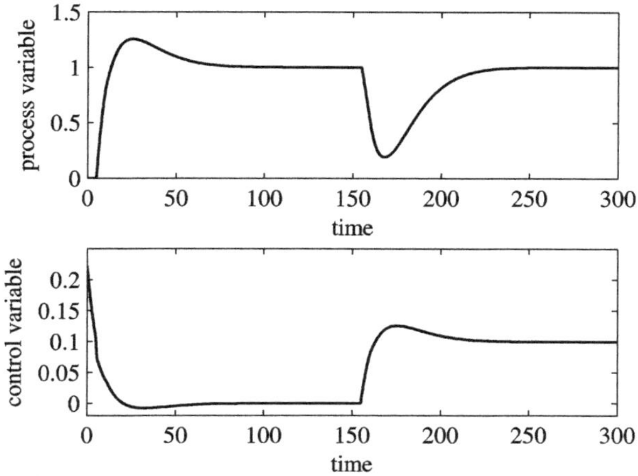

Fig. 8.6 Results obtained by applying the modified Smith predictor (perturbed case)

set-point following and load disturbance rejection. An alternative $P_1(s)$ proposed in [9] is

$$P_1(s) = \frac{\tilde{P}(s)}{(Ls+1)}. \tag{8.9}$$

In this case, if $C(s)$ is a PI controller (see (8.2)), the transfer function between the set-point r and the process variable y is

$$\begin{aligned} H(s) = \frac{Y(s)}{R(s)} &= \frac{\frac{C(s)P(s)}{1+C(s)(P_1(s)-P(s))}}{1+\frac{C(s)P(s)}{1+C(s)(P_1(s)-P(s))}} \\ &= \frac{KK_p\left(s+\frac{1}{T_i}\right)\left(s+\frac{1}{L}\right)e^{-Ls}}{s^3+\frac{1}{L}s^2+\frac{KK_p}{L}s+\frac{KK_p}{T_iL}}, \end{aligned} \tag{8.10}$$

while the transfer function between the load disturbance d and the process variable y is

$$\begin{aligned} H_d(s) = \frac{Y(s)}{D(s)} &= \frac{P(s)}{1+\frac{C(s)P(s)}{1+C(s)(P_1(s)-P(s))}} \\ &= \frac{K\left((1+Ls)s^2+KK_p\left(s+\frac{1}{T_i}\right)(1-(1+Ls)e^{-Ls})\right)e^{-Ls}}{s\left(Ls^3+s^2+KK_ps+\frac{KK_p}{T_i}\right)}. \end{aligned} \tag{8.11}$$

In case a PID controller is employed, namely,

$$C(s) = K_p\left(1+\frac{1}{T_is}+T_ds\right) \tag{8.12}$$

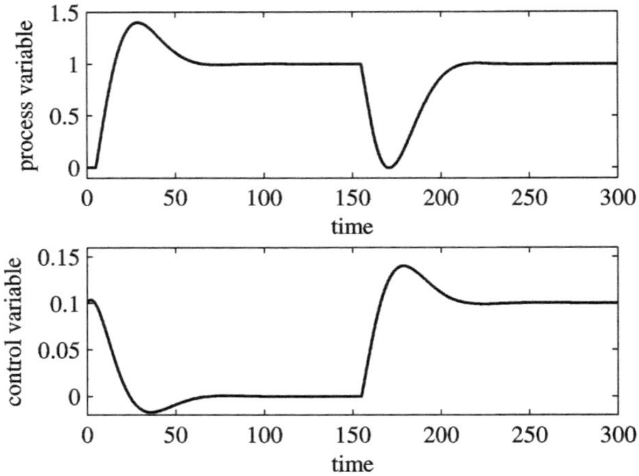

Fig. 8.7 Results obtained by applying the modified Smith predictor with the alternative choice for $P_1(s)$ (nominal case)

it results in

$$H(s) = \frac{Y(s)}{R(s)} = \frac{KK_p s\left(s+\frac{1}{L}\right)\left(T_d s^2 + s + \frac{1}{T_i}\right)e^{-Ls}}{s^3 + \frac{1+KK_pT_d}{L}s^2 + \frac{KK_p}{L}s + \frac{KK_p}{T_iL}} \tag{8.13}$$

and

$$\begin{aligned} H_d(s) &= \frac{Y(s)}{D(s)} \\ &= \frac{K\left((1+Ls)s^2 + KK_p\left(T_d s^2 + s + \frac{1}{T_i}\right)(1-(1+Ls)e^{-Ls})\right)e^{-Ls}}{s\left(Ls^3 + (1+KK_pT_d)s^2 + KK_p s + \frac{KK_p}{T_i}\right)}. \end{aligned} \tag{8.14}$$

Also in this case it can be easily determined that the controller is capable to provide a null steady-state error in the presence of a constant load disturbance. As an illustrative example, consider again process (8.5) and a PI controller (as in Section 8.1) with $K_p = 0.1$ and $T_i = 27$. The response of a unit step set-point signal and of a step load disturbance of amplitude $d = -0.1$ is plotted in Figure 8.7. If there is a 10% estimation error in the dead time (namely, the value of the dead time term employed in the controller is $L = 5.5$), then the result obtained is shown in Figure 8.8.

8.3 Aström–Hang–Lim Modified Smith Predictor

The new Smith predictor devised by Aström, Hang, and Lim [9] has received considerable attention in the last years. Its rationale and the developments that has been proposed are presented hereafter.

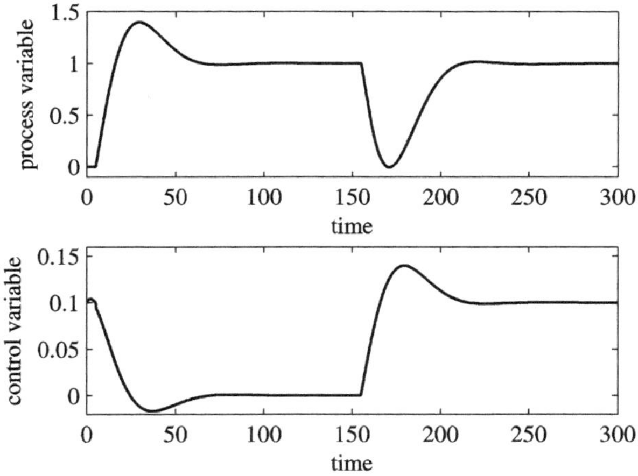

Fig. 8.8 Results obtained by applying the modified Smith predictor with the alternative choice for $P_1(s)$ (perturbed case)

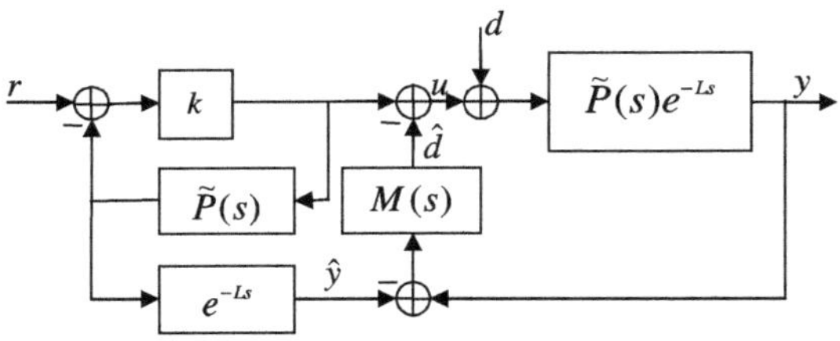

Fig. 8.9 Block diagram of the Aström–Hang–Lim modified Smith predictor

8.3.1 Methodology

The control scheme is shown in Figure 8.9, where it appears that a key role is played by the transfer function $M(s)$ whose output represents an estimate of the load disturbance d. The transfer function between the set-point and the process variable is

$$H(s) = \frac{Y(s)}{R(s)} = \frac{kKe^{-Ls}}{s+kK}\,\frac{1+M(s)\frac{K}{s}e^{-Ls}}{1+M(s)\frac{K}{s}e^{-Ls}} = \frac{kKe^{-Ls}}{s+kK}, \tag{8.15}$$

and it is therefore independent of the $M(s)$ block, which can be designed by considering the load disturbance rejection task only. By noting that the transfer function between the load disturbance and the process output is

$$H_d(s) = \frac{Y(s)}{D(s)} = \frac{\frac{K}{s}e^{-Ls}}{1+M(s)\frac{K}{s}e^{-Ls}}, \tag{8.16}$$

and by considering Expression (8.11), it can be derived that selecting

$$M(s) = \frac{C(s)}{1+\frac{C(s)}{1+C(s)(P_1(s)-P(s))}} = \frac{C(s)}{1+\frac{C(s)}{1+C(s)(\frac{K}{s(1+Ls)}-\frac{K}{s}e^{-Ls})}}, \tag{8.17}$$

the same load response of the modified Smith predictor of Figure 8.4 is obtained. However, the load response can be improved by choosing a different transfer function for $M(s)$, which depends on three design parameters:

$$M(s) = \frac{K_4 + \frac{K_3}{s}}{1+KK_1+\frac{KK_2}{s}+\frac{KK_3}{s^2}-K\left(\frac{K_4}{s}+\frac{K_3}{s^2}\right)e^{-Ls}}, \tag{8.18}$$

where, in order to be $\lim_{s\to 0} H_d(s) = 0$,

$$K_4 = K_2 + K_3 L. \tag{8.19}$$

With this choice, the transfer function between the load disturbance and the process output is

$$H_d(s) = \frac{K((1+KK_1)s^2+KK_2 s+KK_3-K(K_4 s+K_3)e^{-Ls})e^{-Ls}}{s((1+KK_1)s^2+KK_2 s+KK_3)}. \tag{8.20}$$

It can be easily shown that this choice yields a null steady-state error in the presence of a constant load disturbance. Further, it can be noted that, if only the load disturbance is considered, the scheme of Figure 8.9 is equivalent to the scheme of Figure 8.4 if

$$P_1(s) = \tilde{P}(s)\frac{K_1 s^2 + K_2 s + K_3}{K_4 s + K_3}. \tag{8.21}$$

The use of three additional design parameters provides to the two-degree-of-freedom control scheme the capability of improving the performance with respect to the modified Smith predictor control scheme of Figure 8.4.

As an example, consider again process (8.5) and the following values of the controller parameters: $k = 0.5$, $K_1 = 4$, $K_2 = 3$, $K_3 = 0.6$. Then, assuming that the dead time is estimated perfectly (as well as the process gain), it is $K_4 = K_2 + K_3 L = 6$. The results related to the same control task of Section 8.2 are shown in Figure 8.10. It appears that, as expected, the use of a two-degree-of-freedom controller achieves satisfactory performance in both the set-point following and load disturbance rejection tasks. In order to verify the robustness of the controller to modelling uncertainties, a dead time estimation error of 10% is considered, that is, the estimated dead time is $L = 5.5$, and therefore $K_4 = K_2 + K_3 L = 6.3$ (the other controller parameters are the same as in the nominal case). The related results are shown in Figure 8.11.

8.3.2 Robust Tuning Method

The control scheme proposed by Aström, Hang, and Lim appears to be effective, but its tuning can be difficult because there are four parameters involved apart from k and their physical meaning is not entirely clear.

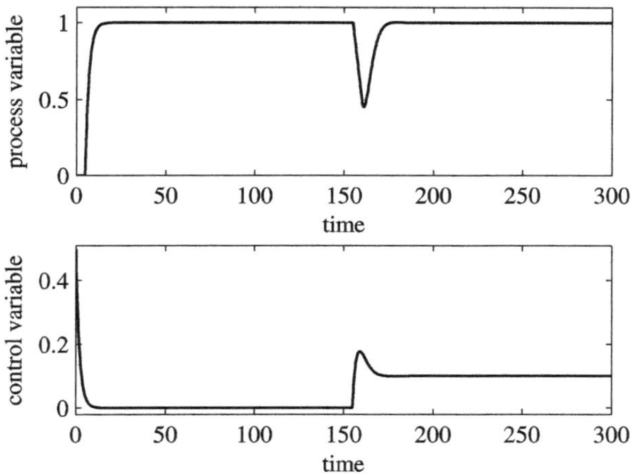

Fig. 8.10 Results obtained by applying the Aström–Hang–Lim modified Smith predictor (nominal case)

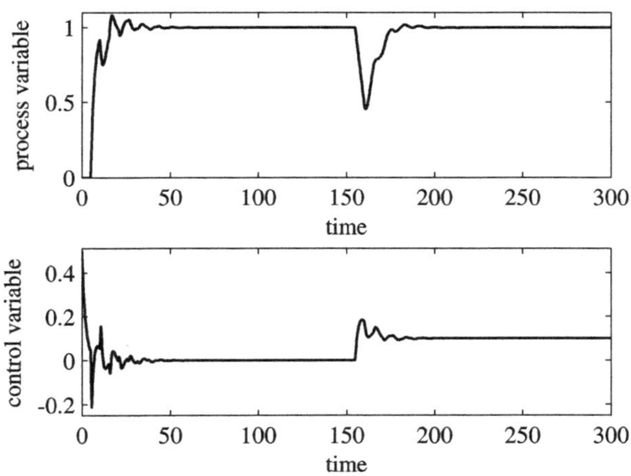

Fig. 8.11 Results obtained by applying the Aström–Hang–Lim modified Smith predictor (perturbed case)

In order to provide an effective help in this context, a robust tuning method has been proposed in [30]. In fact, the block $M(s)$ can be realised as shown in Figure 8.12, where

$$M_0(s) = \frac{KK_4 s + KK_3}{(1+KK_1)\,s^2 + KK_2 s + KK_3}. \tag{8.22}$$

This expression can be rewritten as

$$M_0(s) = \frac{a_3 s + 1}{a_1 s^2 + a_2 s + 1} \tag{8.23}$$

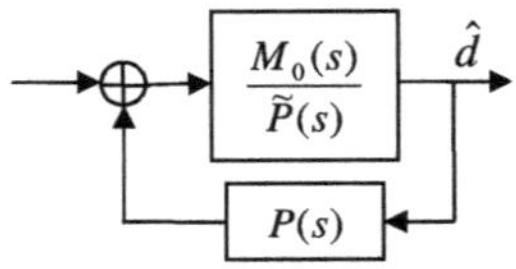

Fig. 8.12 Equivalent representation of the $M(s)$ block

with

$$a_1 = \frac{1+KK_1}{KK_3}, \qquad a_2 = \frac{K_2}{K_3}, \qquad a_3 = \frac{K_4}{K_3}. \tag{8.24}$$

Hence, the transfer function $H_d(s)$ (see (8.20)) can be written as

$$H_d(s) = \left(1 - \frac{M_0(s)}{\tilde{P}(s)} P(s)\right) P(s). \tag{8.25}$$

By taking into account Expressions (8.23) and (8.24), in order to obtain a null steady-state error when a constant load disturbance occurs, condition (8.19) becomes

$$a_3 = a_2 + L. \tag{8.26}$$

Then, the damping factor $a_2/(2\sqrt{a_1})$ for the second-order system (8.23) is chosen as $1/\sqrt{2}$ so that a fast disturbance response with acceptable overshoot is achieved. Thus,

$$a_2 = \sqrt{2a_1}. \tag{8.27}$$

By considering that the (open-loop) transfer function from d to $\hat{d}$ is (see Figure 8.12)

$$W(s) := P(s)M(s) = P(s)\frac{\frac{M_0(s)}{\tilde{P}(s)}}{1 - \frac{M_0(s)}{\tilde{P}(s)} P(s)} = \frac{M_0(s)e^{-Ls}}{1 - M_0(s)e^{-Ls}}, \tag{8.28}$$

we have that the related complementary sensitivity transfer function can be determined as

$$H_0(s) := \frac{W(s)}{1+W(s)} = M_0(s)e^{-Ls}. \tag{8.29}$$

At this point, consider a process model $P_m(s)$ with multiplicative uncertainty, namely,

$$P(s) = P_m(s)\big(1 + \Delta(s)\big), \tag{8.30}$$

where

$$P_m(s) = \frac{K_m}{s} e^{-L_m s}, \tag{8.31}$$

and assume that the uncertainty is only for the dead time (this is the most critical case), namely ($K = K_m$),

$$P(s) = \frac{K}{s} e^{-(L_m + \Delta L)s} = P_m(s)e^{-\Delta L s}. \tag{8.32}$$

The uncertainty can be therefore seen as a multiplicative uncertainty where

$$\Delta(s) = e^{-\Delta L s} - 1. \tag{8.33}$$

The robust stability criterion for multiplicative uncertainty is given by the following expression [22]:

$$\left\| \Delta(s) H_0(s) \right\|_\infty < 1 \tag{8.34}$$

which, by taking into account (8.29) can be rewritten as

$$\left| M_0(j\omega) \right| < \frac{1}{|\Delta(j\omega)|} \quad \forall \omega, \tag{8.35}$$

that is (see (8.23) and (8.33)),

$$\left| \frac{j a_3 \omega + 1}{1 - a_1 \omega^2 + j a_2 \omega} \right| < \frac{1}{|e^{-j\omega \Delta L} - 1|} \quad \forall \omega. \tag{8.36}$$

Now, consider the delay margin D_m, *i.e.*, the maximum delay error that the system can tolerate before becoming unstable. According to the definition of phase margin, it is

$$D_m = \frac{\phi_m}{\omega_c} \frac{\pi}{180}, \tag{8.37}$$

where ω_c is the gain crossover frequency. If $\Delta L = D_m$, then it is

$$\left\| \Delta(j\omega_c) H_0(j\omega_c) \right\| = \left\| \Delta(j\omega_c) M_0(j\omega_c) \right\| = 1. \tag{8.38}$$

Hence, by approximating the dead time term with a first-order Taylor series expansion, *i.e.*, $e^{-\Delta L} \cong 1 - \Delta L s$, it can be deduced that

$$\frac{1 + a_3^2 \omega_c^2}{(1 - a_1 \omega_c^2)^2 + a_2^2 \omega_c^2} = \frac{1}{D_m^2 \omega_c^2}. \tag{8.39}$$

Now, consider the phase margin $\phi_m = 180/\pi \cong 57.3°$, which is a sensible value. In this case, $D_m = 1/\omega_c$ (see (8.37)), and, by considering also Expression (8.27), Equation (8.39) becomes

$$\frac{1 + \frac{a_3^2}{D_m^2}}{1 + \frac{a_1^2}{D_m^4}} = 1, \tag{8.40}$$

that is,

$$D_m = \frac{a_1}{a_3}. \tag{8.41}$$

It is sensible to choose a delay margin proportional to the maximum delay uncertainty $\Delta L_{\max}$. By considering $D_m = 2\Delta L_{\max}$, it results

$$a_1 = 2 a_3 \Delta L_{\max}. \tag{8.42}$$

Thus, by considering (8.26), (8.27), and (8.42), the values of the parameters of $M_0(s)$ can be obtained as

$$a_2 = 2\left(\Delta L_{\max} + \sqrt{\Delta L_{\max}^2 + L_m \Delta L_{\max}}\right),$$
$$a_1 = \frac{a_2^2}{2}, \tag{8.43}$$
$$a_3 = a_2 + L_m.$$

In order to tune the parameters of the $M(s)$ transfer function (8.18), it is sufficient to select

$$K_3 = 1, \tag{8.44}$$

and then from (8.24) it can be deduced that

$$K_2 = 2\left(\Delta L_{\max} + \sqrt{\Delta L_{\max}^2 + L_m \Delta L_{\max}}\right), \tag{8.45}$$
$$K_1 = \frac{1}{2}K_2^2 - \frac{1}{K}, \tag{8.46}$$
$$K_4 = K_2 + L_m. \tag{8.47}$$

As an illustrative example, consider the true process

$$P(s) = \frac{1}{s}e^{-5s} \tag{8.48}$$

and consider the following process model:

$$P_m(s) = \frac{1}{s}e^{-5.5s}, \tag{8.49}$$

that is, $K = K_m = 1$, $L = 5$, and $L_m = 5.5$. Then, by selecting $\Delta L_{\max} = 0.5$, formulae (8.44)–(8.47) give $K_1 = 8.96$, $K_2 = 4.46$, $K_3 = 1$, and $K_4 = 9.96$. In addition, the gain $k = 0.5$ has been selected. The set-point and load disturbance step responses are shown in Figure 8.13, where the effectiveness of the tuning procedure appears.

8.3.3 Simplified Tuning Method

As an alternative to the robust tuning method presented in Section 8.3.2, the simple approach proposed in [153] can be considered. It consists in representing the $M(s)$ block as shown in Figure 8.14.

In this case, the disturbance response is given by the transfer function (see (8.20) and (8.25))

$$H_d(s) = \left(1 - sM_0(s)P(s)\right)P(s) = \left(1 - M_0(s)Ke^{-Ls}\right)\frac{K}{s}e^{-Ls}. \tag{8.50}$$

In order to obtain a null steady-state error in response to a constant load disturbance, the condition

$$\lim_{s\to 0} H_d(s) = 0 \tag{8.51}$$

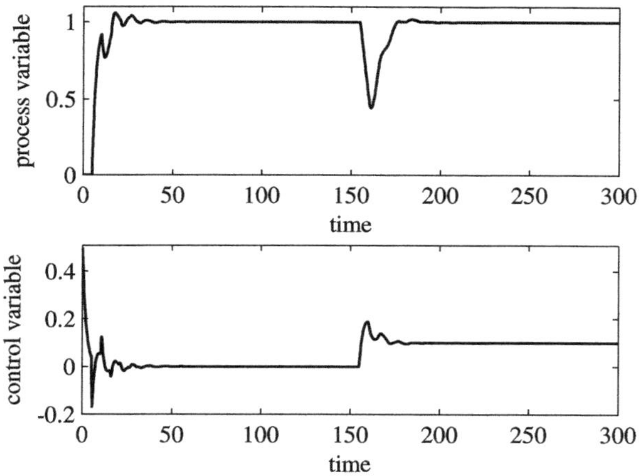

Fig. 8.13 Results obtained by applying the robust tuning method to the Aström–Hang–Lim modified Smith predictor (perturbed case)

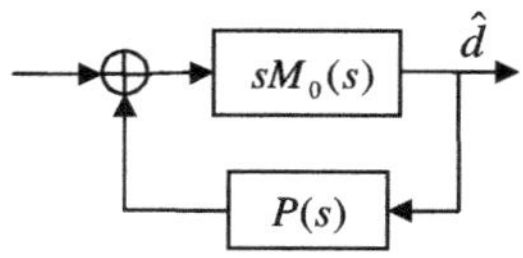

Fig. 8.14 Another equivalent representation of the $M(s)$ block

is equivalent to the condition that zero $s=0$ in $1-M_0(s)Ke^{-Ls}$ has a multiplicity equal to two. This implies that

$$\lim_{s\to 0}\left(1-M_0(s)Ke^{-Ls}\right)=0 \tag{8.52}$$

and

$$\lim_{s\to 0}\frac{d}{ds}\left(1-M_0(s)Ke^{-Ls}\right)=0. \tag{8.53}$$

Condition (8.52) yields

$$M_0(0)=\frac{1}{K}, \tag{8.54}$$

while condition (8.53) yields

$$2LM_0(0)-\frac{L}{K}-\dot{M}_0(0)=0. \tag{8.55}$$

Thus, $M_0(s)$ should satisfy (8.54) and (8.55) at the same time. A possible solution is to consider

$$M_0(s)=\frac{1}{K}\frac{\alpha s+1}{(\lambda s+1)^2}, \tag{8.56}$$

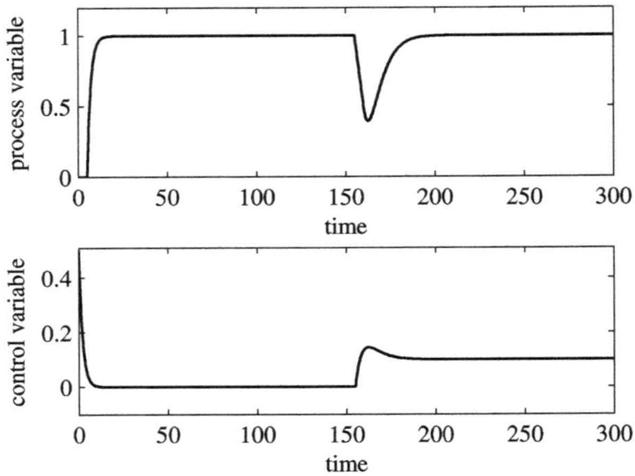

Fig. 8.15 Results obtained by applying the simplified tuning method to the Aström–Hang–Lim modified Smith predictor (nominal case)

where, by considering (8.55), it has to be

$$\alpha = 2\lambda + L. \tag{8.57}$$

Thus,

$$M_0(s) = \frac{1}{K}\frac{(2\lambda + L)s + 1}{(\lambda s + 1)^2} \tag{8.58}$$

and

$$H_d(s) = \frac{(\lambda s + 1)^2 - ((2\lambda + L)s + 1)e^{-Ls}}{(\lambda s + 1)^2}\frac{K}{s}e^{-Ls}. \tag{8.59}$$

It appears that the disturbance response depends on the design parameter λ which can be tuned in order to meet the desired specifications. In particular, it has to be noted that $M_0(s)$ is, in fact, a low-pass filter whose bandwidth is a monotonic function of λ. In any case, it is difficult to address the transfer function (8.59) analytically, and therefore a rule of thumb, which considers modelling uncertainties, has been established [153]. It is suggested to select λ in the interval between $0.5L$ and $1.5L$.

As an illustrative example, process (8.5) is considered again. By selecting $k = 0.5$ for the set-point response and $\lambda = L = 5$ for the load disturbance rejection, the obtained results are shown in Figure 8.15. If the dead time term is estimated as $L = 5.5$ (and the value of $\lambda = 5.5$ is selected accordingly), the obtained results are those shown in Figure 8.16. Despite the simplicity of the technique, a high performance is achieved.

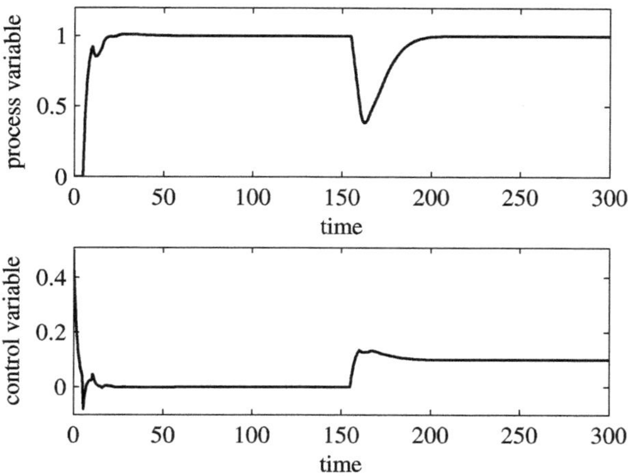

Fig. 8.16 Results obtained by applying the simplified tuning method to the Aström–Hang–Lim modified Smith predictor (perturbed case)

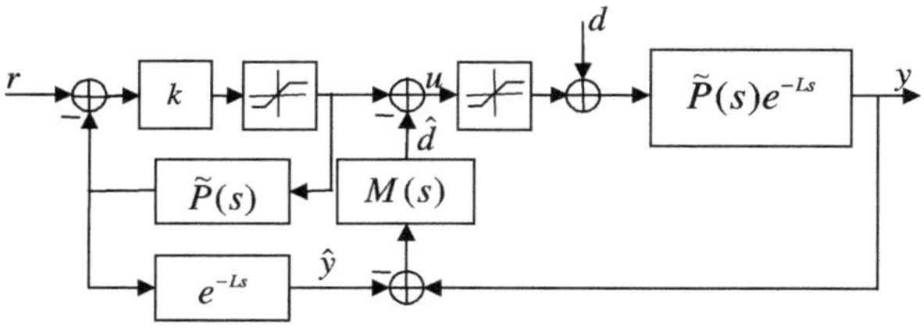

Fig. 8.17 Anti-windup compensation scheme for the modified Smith predictor

8.3.4 Anti-windup Compensation

In the presence of an actuator saturation, the performance obtained by the modified Smith predictor can be decreased by the windup phenomenon (see Section 2.1.2). In order to solve this problem, a very simple anti-windup compensation strategy has been proposed in [28]. An additional saturation block is placed in the control scheme, just after the gain k, as shown in Figure 8.17. This saturation block should describe the actuator constraint acting on the process input (which has also been inserted in the scheme of Figure 8.17). The added saturation block aims at predicting (and therefore compensating) the adverse effects of the process input saturation.

As an illustrative example of the effectiveness of this simple technique, the same nominal case of Section 8.3.3 is considered, where an input saturation limit of 0.2 has been implemented. The set-point unit step response without the anti-windup compensation in shown in Figure 8.18. Only the set-point response has been considered because it is the one that is mainly affected by the windup phenomenon. It is evident that the performance is worsen by the presence of the saturation, as an overshoot appears and the settling time is increased. Conversely, by applying the anti-

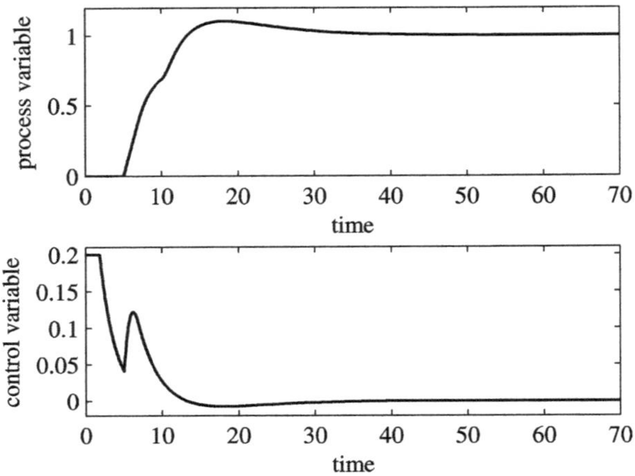

Fig. 8.18 Results obtained by considering input saturation

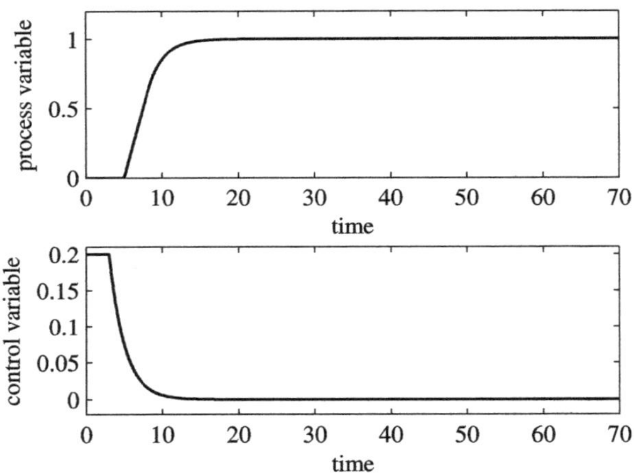

Fig. 8.19 Results obtained by considering input saturation and an anti-windup technique

windup technique (see Figure 8.19), the set-point response is smooth (and without overshoot) as in the case where the input saturation is not involved (see Figure 8.15). Similar results are obtained also in the presence of modelling uncertainties.

8.4 Matausek–Micic Modified Smith Predictor

A simple and straightforward modification of the original Smith predictor to cope with integral processes has been presented in [69]. The basic scheme and a further improvement are explained in the next subsections.

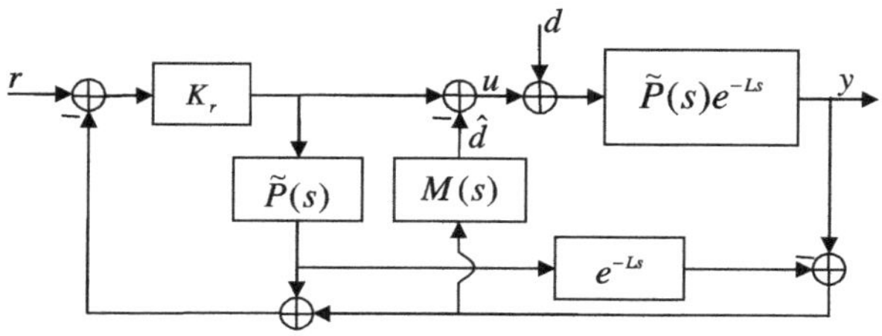

Fig. 8.20 Block diagram of the Matausek–Micic modified Smith predictor

8.4.1 Basic Scheme

The modification consists in adding an additional feedback loop from the difference of the process output and the model output to the control variable, as it is shown in Figure 8.20, where $M(s)$ is a simple gain, *i.e.*,

$$M(s) = K_0. \tag{8.60}$$

Actually, $\hat{d}$ is evidently an estimate of the load disturbance d. In the nominal case, the set-point and load disturbance response is given by (note that it is assumed that the process is described by an IPDT model (8.1))

$$\begin{aligned} Y(s) = & \frac{K_r K e^{-Ls}}{s + K K_r} R(s) \\ & + \frac{K(s + K K_r(1 - e^{-Ls}))e^{-Ls}}{(s + K K_r)(s + K_0 K e^{-Ls})} D(s). \end{aligned} \tag{8.61}$$

It can be noted that

$$\lim_{s \to 0} \frac{K(s + K K_r(1 - e^{-Ls}))e^{-Ls}}{(s + K K_r)(s + K_0 K e^{-Ls})} = 0, \tag{8.62}$$

namely, a null steady-state error is ensured in the presence of a constant load disturbance. Moreover, the scheme has a two-degree-of-freedom structure, because only the load disturbance compensation performance depends on the design parameter K_0.

In order to provide rules for the tuning of the controller (*i.e.*, for the determination of the values of K_0 and K_r), the following considerations can be done. From (8.74) it is apparent that the stability of the modified Smith predictor control system depends on the roots of the characteristic equation

$$(s + K K_r)\left(s + K_0 K e^{-Ls}\right) = 0. \tag{8.63}$$

While an analysis of the roots of the equation

$$s + K_0 K e^{-Ls} = 0 \tag{8.64}$$

can be performed by applying a root locus technique, for the purpose of tuning, it is sufficient to determine the ultimate gain K_{0u}. In this context, Equation (8.64) can be rewritten as

$$1 + W(s) = 0, \tag{8.65}$$

where, evidently,

$$W(s) := \frac{K_0 K}{s} e^{-Ls}. \tag{8.66}$$

Then, the ultimate gain K_{0u} can be determined by setting the phase margin $\phi_m = \pi + \arg\{W(j\omega_c)\}$ of $W(s)$ to zero, that is, by solving the following system:

$$\pi + \arg\{W(j\omega_c)\} = 0, \tag{8.67}$$

$$\left|W(j\omega_c)\right| = 1. \tag{8.68}$$

It yields

$$K_{0u} = \frac{\pi}{2KL}. \tag{8.69}$$

A sensible value of K_0 is therefore given by (see Section 7.4.2)

$$K_0 = \frac{\frac{\pi}{2} - \phi_m}{KL}, \tag{8.70}$$

that is,

$$K_0 = \frac{1}{2KL} \tag{8.71}$$

if a phase margin of $W(s)$ equal to $\phi_m = 61.3°$ is selected.

Regarding the choice of the parameter K_r, it is worth rewriting it as

$$K_r = \frac{1}{KT_r} \tag{8.72}$$

because, in this way, it is clear from Expression (8.61) that T_r is the time constant of the closed-loop set-point step response. Thus, T_r has a clear physical meaning and can be selected by considering the trade-off between aggressiveness and robustness. In any case, a rigorous robust tuning procedure has been presented in [43].

As an illustrative example, the same control task of the previous subsections is considered, namely, a unit step signal is applied to the set-point at time $t = 0$, and a load step disturbance of amplitude –0.1 is applied at time $t = 150$ to process (8.5). The result obtained by selecting $T_r = 2$ (*i.e.*, $K_r = 0.5$) in the nominal case (that is, $K_0 = 1/(2KL) = 0.1$) is shown in Figure 8.21. If the perturbed case in considered, that is, the dead time is estimated as $L = 5.5$ and therefore $K_0 = 1/(2KL) = 0.909$, the resulting process and control variable are those plotted in Figure 8.22.

It is worth stressing at this point that, in case the value of the dead time is not very significant (with respect to the dominant time constant of the process), a similar approach based on the IMC concept can be employed (note that it can be applied to SOIPDT processes) [46, 48].

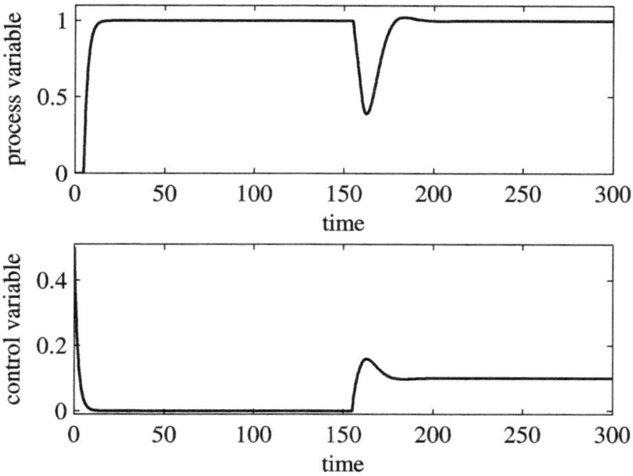

Fig. 8.21 Results obtained by applying the Matausek–Micic modified Smith predictor (nominal case)

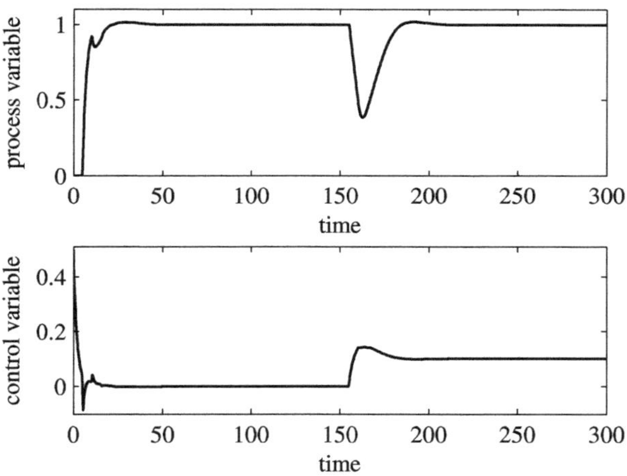

Fig. 8.22 Results obtained by applying the Matausek–Micic modified Smith predictor (perturbed case)

8.4.2 Improvement

An improvement of the previous scheme has been presented in [70]. It consists in choosing $M(s)$ in the control scheme of Figure 8.20 as

$$M(s) = \frac{K_0\,(T_d s + 1)}{T_f s + 1}, \qquad T_f = \frac{T_d}{10}, \tag{8.73}$$

that is, a (filtered) PD structure is employed instead of a proportional gain. In this case, the set-point and load disturbance response is given by

$$Y(s) = \frac{K_r K e^{-Ls}}{s + K K_r} R(s) + \frac{K(s + K K_r(1 - e^{-Ls}))e^{-Ls}}{(s + K K_r)\left(s + \frac{K_0 K (T_d s+1)}{T_f s+1} e^{-Ls}\right)} D(s), \tag{8.74}$$

and therefore the stability of the modified Smith predictor control system depends on the roots of the characteristic equation

$$(s + K K_r)\left(s + \frac{K_0 K\,(T_d s + 1)}{T_f s + 1} e^{-Ls}\right) = 0. \tag{8.75}$$

By following the same approach of the previous subsection, the following characteristic equation can be analysed:

$$1 + W(s) = 0, \tag{8.76}$$

where

$$W(s) := \frac{K_0 K\,(T_d s + 1)}{s(T_f s + 1)} e^{-Ls}. \tag{8.77}$$

Then, by choosing the value of the derivative time constant T_d to be proportional to the dead time L, that is,

$$T_d = \alpha L, \quad 0 \le \alpha < 1, \tag{8.78}$$

one can obtain the solution of Equation (8.67) after having neglected the first-order filter in $M(s)$ (because $T_f = T_d/10$) and after having exploited the following approximation:

$$\arctan(\alpha L \omega_c) \cong \alpha L \omega_c. \tag{8.79}$$

It results to be

$$\omega_c = \frac{\frac{\pi}{2} - \phi_m}{(1 - \alpha) L}, \quad 0 < \phi_m < \frac{\pi}{2}. \tag{8.80}$$

By assuming that

$$0 \le K K_0 \alpha L < 1, \tag{8.81}$$

from (8.68) it can be derived that

$$\omega_c = \frac{K K_0}{\sqrt{1 - (K K_0 \alpha L)^2}}, \tag{8.82}$$

and, finally, by considering (8.80) and (8.82), the gain K_0 can be obtained as

$$K_0 = \frac{\frac{\pi}{2} - \phi_m}{K L \sqrt{(1 - \alpha)^2 + \left(\frac{\pi}{2} - \phi_m\right)^2 \alpha^2}}. \tag{8.83}$$

It appears from (8.78) and (8.83) that the choice of α and ϕ_m allows the determination of T_d and K_0. Note also that the choice $\alpha = 0$ yields the case of the

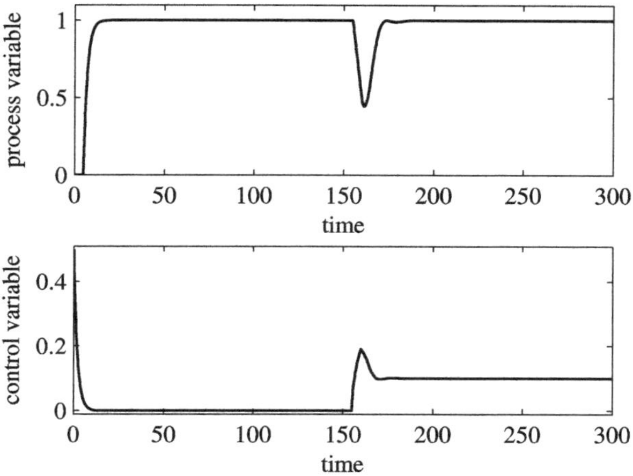

Fig. 8.23 Results obtained by applying the improved Matausek–Micic modified Smith predictor (nominal case)

previous subsection (see (8.70)). In [70], based on a large number of simulations, it is suggested to select

$$\alpha = 0.4, \qquad \phi_m = 64^\circ, \tag{8.84}$$

while the same considerations done in the previous subsection can be applied for the tuning of K_r (see (8.72)).

As an illustrative example, consider the same control task of the previous subsection. In the nominal case, the application of the tuning rules (8.84) gives $K_0 = 0.145$ and $T_d = 2$ ($T_f = 0.2$), while the same value $T_r = 2$ (*i.e.*, $K_r = 0.5$) is selected. The resulting process and control variables are plotted in Figure 8.23. Conversely, if the dead time is estimated as $L = 5.5$, the tuning formula (8.84) gives $K_0 = 0.132$ and $T_d = 2.2$ ($T_f = 0.22$), and the corresponding results are shown in Figure 8.24.

It is worth stressing that, in order to cope with the possible occurrence of the windup phenomenon in the presence of actuator saturation, the same method described in Section 8.3.4 can be applied also to the scheme of Figure 8.20 [29]. Indeed, it is sufficient to insert a model of the actuator saturation just after the gain K_r in order to compensate the windup effect efficiently.

Finally, note that the use of a sliding mode controller instead of the proportional controller K_r has been proposed in [13].

8.5 Normey-Rico–Camacho Modified Smith Predictor

The two-degree-of-freedom control scheme proposed in [78] has received a significant attention in the last few years, as it is shown in the following subsections.

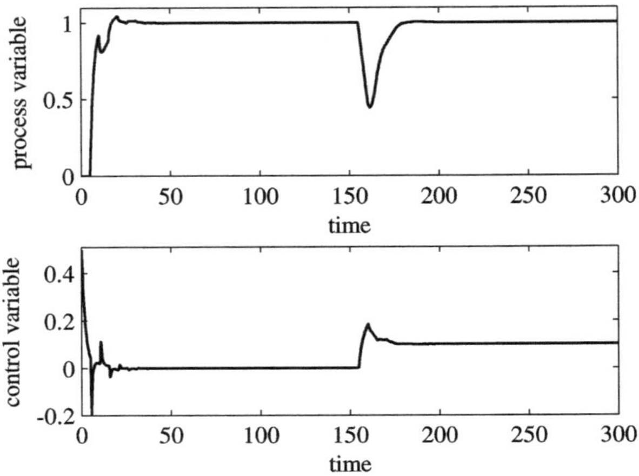

Fig. 8.24 Results obtained by applying the improved Matausek–Micic modified Smith predictor (perturbed case)

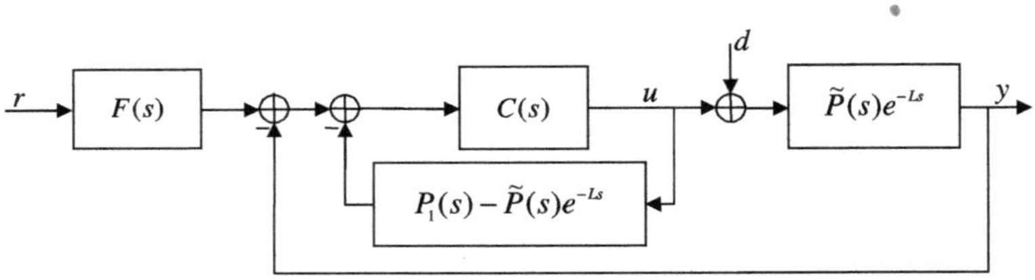

Fig. 8.25 Block diagram of the Normey-Rico–Camacho modified Smith predictor

8.5.1 Control Scheme

The Normey-Rico and Camacho scheme consists of a simple modification of the scheme of Figure 8.4 and aims at improving the set-point response. In particular, a filter $F(s)$ is applied to the set-point as in the typical two-degree-of-freedom control schemes (see Figure 2.3). The overall control scheme is shown in Figure 8.25.

In this case, the following choices are made (see (8.6) for the choice of $P_1(s)$):

$$C(s) = K_p\left(1 + \frac{1}{T_i s}\right), \tag{8.85}$$

$$P_1(s) = K\frac{1 - Ls}{s}. \tag{8.86}$$

Obviously, the transfer function between the load disturbance and the process output does not depend on the prefilter $F(s)$, while the transfer function from the set-point to the process output is (see (8.7))

$$H(s) = \frac{F(s)C(s)P(s)}{1 + C(s)P_1(s)}. \tag{8.87}$$

In the scheme of Figure 8.4, if a high performance is obtained in the load disturbance rejection task, a significant overshoot results in the set-point step response because of the zero of the primary controller $C(s)$. Thus, the prefilter $F(s)$ can be employed effectively to cancel this zero by selecting

$$F(s) = \frac{1+\alpha T_i s}{1+T_i s}, \quad \alpha < 1. \tag{8.88}$$

8.5.2 Robust Tuning

The control scheme of Figure 8.25 requires the tuning of three parameters: K_p, T_i, and α. A typical goal of a design procedure is to obtain a critically damped closed-loop system (in order to avoid overshoots), which is as fast as possible. By considering the characteristic equation of the closed-loop system (see (8.7))

$$T_i(1-KK_pL)s^2 + KK_p(T_i - L)s + KK_p = 0, \tag{8.89}$$

it can be derived that, in order for the closed-loop system to have a double pole in $s = -1/T_0$, the following relations have to be verified:

$$T_i = 2T_0 + L, \tag{8.90}$$

$$K_p = \frac{2T_0 + L}{K(T_0+L)^2}. \tag{8.91}$$

Then, the value $\alpha = 0.4$ is suggested after having performed a large number of simulations.

At this point, the value of T_0 has to be selected. For this purpose, a robustness analysis can be performed. By considering that a nominal process model denoted as $P_m(s) = K_m e^{-L_m s}/s$ is related to the real process $P(s)$ as

$$P(s) = P_m(s) + \Delta P(s), \tag{8.92}$$

the norm-bound uncertainty region $|DP|$ is determined such that the closed-loop stability is maintained if $|\Delta P| < |DP|$. The characteristic equation of the closed-loop system in the presence of uncertainty is

$$1 + C(s)\big(P_1(s) + \Delta P(s)\big), \tag{8.93}$$

and, therefore, by solving for $\Delta P(s)$, the norm-bound uncertainty can be determined as

$$|DP| = \frac{|1 + C(j\omega)P_1(j\omega)|}{|C(j\omega)|} = \frac{|K(1+j\omega T_0)^2|}{|j\omega(1+j\omega(2T_0+L))|}, \quad \omega > 0. \tag{8.94}$$

The minimum of the normalised norm-bound uncertainty, denoted as $\delta_{\min}$, can be obtained as

$$\delta_{\min} = \min\big|DP \cdot K^{-1}\big| = \big|T_0^2(2T_0+L)^{-1}\big|, \tag{8.95}$$

and, therefore,

$$T_0 = \delta_{\min}\left(1+\sqrt{1+\frac{L}{\delta_{\min}}}\right). \tag{8.96}$$

Hence, the value of T_0 can be selected based on value $\delta_{\min}$ which has to be related to the uncertainties of the process. Typically, for processes with dominant dead time, the robustness issue can be addressed by considering only dead-time estimation errors [80]. In this case, the normalised uncertainty can be described as

$$\overline{\Delta P}(s) = \frac{P(s)-P_m(s)}{K} = \frac{1}{s}e^{-Ls}\left(1-e^{-\Delta Ls}\right), \tag{8.97}$$

where

$$\Delta L = L_m - L. \tag{8.98}$$

Then, by approximating $e^{-\Delta Ls}$ as $1-\Delta Ls$, Equation (8.97) can be rewritten as

$$\overline{\Delta P}(s) = e^{-Ls}\Delta L. \tag{8.99}$$

The modulus of the previous expression for $s = j\omega$ is evidently constant, and its value is $|\Delta L|$. Based on the above analysis, the system is robustly stable, in principle, if the value $\delta_{\min} = \Delta L$ is chosen for the selection of T_0 by means of Expression (8.96). However, by taking into account the approximations done and the fact that a robust performance is pursued, it is suggested to select the value of $\delta_{\min}$ between $1.5\Delta L$ and $2\Delta L$ if a small overshoot is required.

As a first illustrative example, process (8.5) is considered without uncertainty, and the value $T_0 = 2$ is selected, for which $T_i = 9$ and $K_p = 0.18$ (see (8.90) and (8.91)). The set-point and load disturbance step responses are shown in Figure 8.26.

As a second example, as in the previous sections, a dead-time estimation error $\Delta L = 0.5$ is considered (namely, the estimated dead time is $L_m = 5.5$, whereas the true dead time is $L = 5$. Thus, the value of $\delta_{\min} = 2\cdot 0.5 = 1$ is chosen, and the consequent value of T_0 is 3.55 according to Expression (8.96). The resulting values of T_i and K_p are therefore 12.6 and 0.154, respectively. The resulting set-point and load disturbance step responses are plotted in Figure 8.27.

8.5.3 Improvement

An improvement of the method described in the previous subsections has been presented in [80]. While the primary PI controller $C(s)$ is chosen as in (8.85) with K_p and T_i given by (8.91) and (8.90), respectively, the filter $F(s)$ is selected, differently from (8.88), as the second-order transfer function

$$F(s) = \frac{(1+T_0 s)^2}{(1+T_i s)(1+T_1 s)}. \tag{8.100}$$

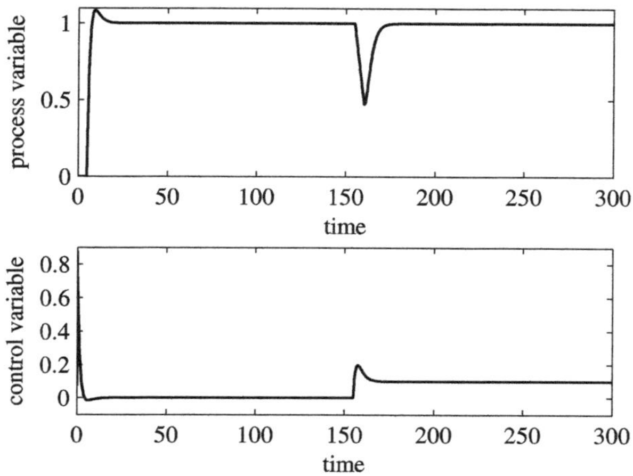

Fig. 8.26 Results obtained by applying the Normey-Rico–Camacho modified Smith predictor (nominal case)

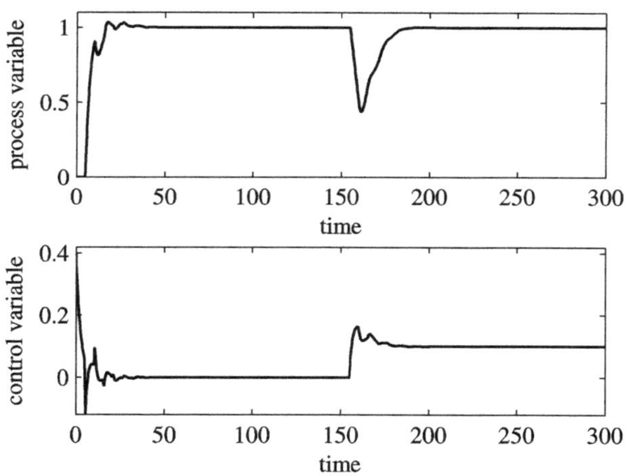

Fig. 8.27 Results obtained by applying the Normey-Rico–Camacho modified Smith predictor (perturbed case)

In this case, the closed-loop transfer function between the set-point and the process output is

$$H(s) = \frac{Y(s)}{R(s)} = \frac{1}{T_1 s + 1} e^{-Ls}, \tag{8.101}$$

while that between the load disturbance and the process output is

$$H_d(s) = \frac{K}{s} e^{-Ls} \left(1 - \frac{(2T_0 + L)s + 1}{(T_0 s + 1)^2} e^{-Ls} \right). \tag{8.102}$$

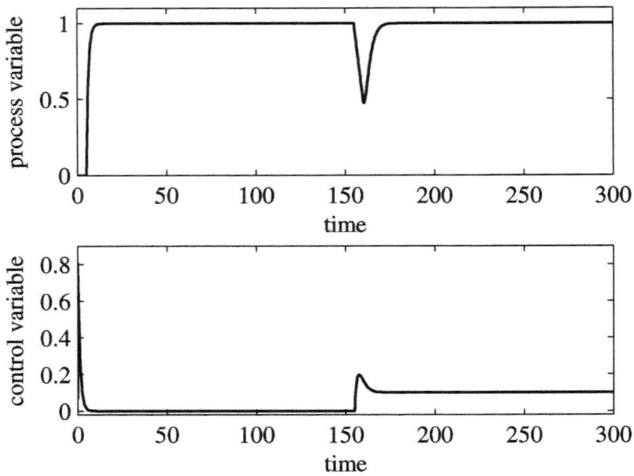

Fig. 8.28 Results obtained by applying the improved Normey-Rico–Camacho modified Smith predictor (nominal case)

It appears that T_1 determines the set-point response, while T_0 determines the load disturbance response, and therefore the two control tasks are fully decoupled. By taking into account the robustness considerations done in the previous subsections, the same tuning rule (8.96) can be employed for T_0, while for T_1, it is suggested to select

$$T_1 = 0.6T_0 + T_e, \tag{8.103}$$

where T_e is the equivalent time constant of the non-integral part of the process (which has been assumed to be zero in the previous analysis).

The results obtained by employing the improved scheme for the nominal case (where the same control task of the previous subsection has been considered) are shown in Figure 8.28. Note that, again, $T_0 = 2$ has been chosen, and, as a consequence, $T_i = 9$, $K_p = 0.18$, and $T_1 = 0.6T_0 = 1.2$. On the contrary, when the dead time is erroneously estimated as $L = 5.5$, the design parameters are determined as $T_0 = 3.55$, $T_i = 12.6$, $K_p = 0.154$, and $T_1 = 0.6T_0 = 2.13$, and the results obtained are plotted in Figure 8.29.

8.5.4 An Alternative Approach

The design method proposed by Normey-Rico and Camacho can be derived also by considering an Internal Model Control approach [76] (see Section 2.3.2.1), as pointed out in [155]. Indeed, the control scheme of Figure 8.25 is equivalent to the IMC scheme of Figure 2.16, where

$$Q(s) = \frac{C(s)}{1 + C(s)P_1(s)}. \tag{8.104}$$

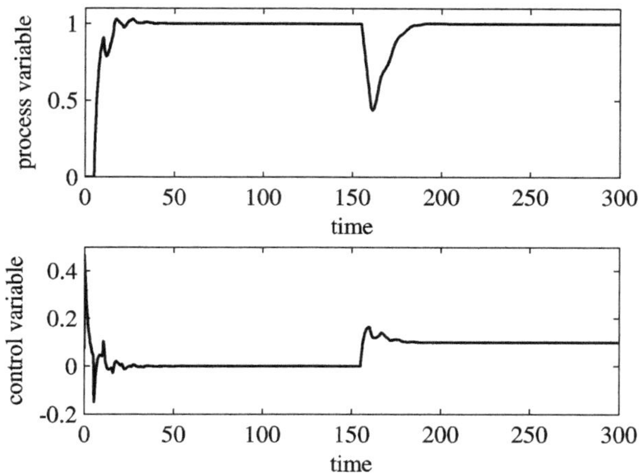

Fig. 8.29 Results obtained by applying the improved Normey-Rico–Camacho modified Smith predictor (perturbed case)

By applying the IMC design procedure (which yields an H_2 optimal controller), it can be obtained that

$$C(s) = \frac{1}{K}\frac{(2\lambda_C + L)s + 1}{(\lambda_C + L)^2 s}, \tag{8.105}$$

where λ_C is the filter time constant (see (2.65)). It can be easily seen that this controller transfer function is equivalent to the transfer function (8.85) with the tuning rules (8.90)–(8.91) if $\lambda_C = T_0$. Then, by considering that the IMC control system transfer function (see Figure 2.17) results to be

$$\frac{(2\lambda_C + L)s + 1}{(\lambda_C s + 1)^2} e^{-Ls}, \tag{8.106}$$

the optimal prefilter transfer function (which minimises the H_2 norm for $\lambda_F \to 0$) can be determined as

$$F(s) = \frac{(\lambda_C s + 1)^2}{((2\lambda_C + L)s + 1)(\lambda_F s + 1)}, \tag{8.107}$$

which is equivalent to the prefilter transfer function (8.100) with again $\lambda_C = T_0$ and $\lambda_F = T_1$ (in addition to $T_i = 2T_0 + L$).

Thus, it can be concluded that the design methodology proposed in the previous subsections is, in fact, optimal. A possible advantage of using an IMC approach to derive the same result is that it is possible to extend it easily to processes with double integrators and dead time [155].

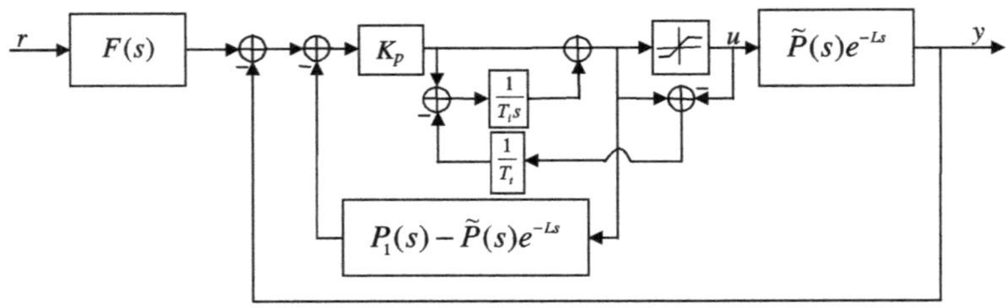

Fig. 8.30 Block diagram of the Normey-Rico–Camacho modified Smith predictor with anti-windup compensation

8.5.5 Anti-windup Strategy

As an anti-windup strategy, the back-calculation technique explained in Section 2.1.2 (see Figure 2.4) can be applied straightforwardly to the modified Smith predictor of Figure 8.25 [152]. Indeed, the primary controller $C(s)$ can be decomposed into a proportional and an integral part, and therefore the integral action can be reduced by a term proportional to the saturation level of the actuator. The corresponding control scheme is shown in Figure 8.30, where T_t is the tracking time constant, which is suggested to be selected (based on a stability analysis) as

$$T_t = \left(\frac{\alpha+1}{2\alpha+1}\right)^2, \tag{8.108}$$

where

$$\alpha = \frac{T_0}{L}. \tag{8.109}$$

The effectiveness of the anti-windup compensation technique is shown with the following example, where the same control task of Section 8.3.4 is considered, namely, an actuator saturation limit of 0.2 is applied to the process input. By considering $T_0 = 2$ and $T_1 = 1.2$, as in Figure 8.28, the set-point unit step response without compensation is that shown in Figure 8.31 (note that the nominal case is considered). On the contrary, if the back-calculation technique is applied with $T_t = 0.605$, the result obtained is that shown in Figure 8.32, where it appears that the overshoot and the settling time are greatly reduced.

8.5.6 Comparison with Other Schemes

A simple comparison between different control schemes presented in the previous sections can be made by considering the nominal case [79]. In particular, the scheme of Figure 8.9 with the simplified tuning procedure of Section 8.3.3, the scheme of Figure 8.20 with $M(s) = K_0$ (see Section 8.4.1) and that of Figure 8.25, where $F(s)$ is a second-order system (see Section 8.5.3), are considered hereafter.

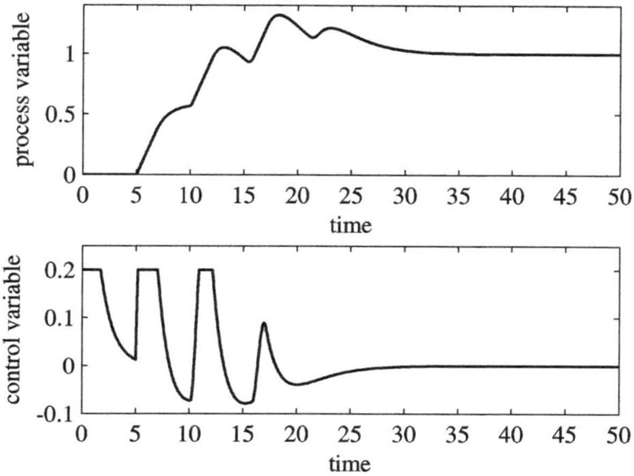

Fig. 8.31 Results obtained by applying the improved Normey-Rico–Camacho modified Smith predictor with actuator saturation and without anti-windup compensation

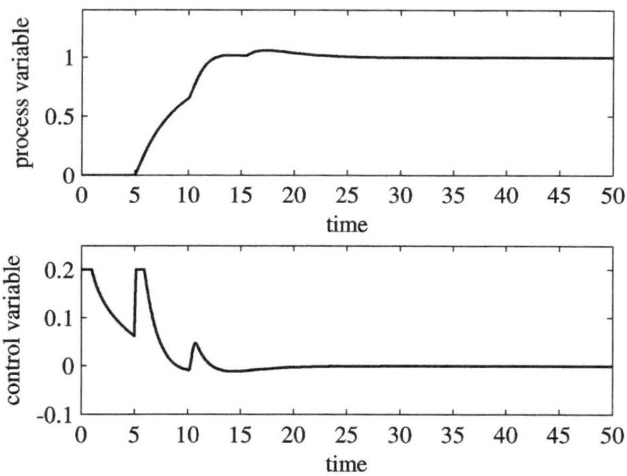

Fig. 8.32 Results obtained by applying the improved Normey-Rico–Camacho modified Smith predictor with actuator saturation and with anti-windup compensation

If the set-point response is considered, it appears that all the methods provide the same performance, because the closed-loop systems have in all the cases a first-order-plus-dead-time transfer function with time constant given by $1/(kK)$ in the Aström–Hang–Lim controller (see (8.15)), by $1/(KK_r)$ in the Matausek–Micic controller (see (8.74)), and by T_1 in the Normey-Rico–Camacho controller (see (8.101)).

Conversely, the transfer function between the load disturbance and the process output is the same in the Aström–Hang–Lim and Normey-Rico–Camacho schemes (provided that $\lambda = T_0$, see (8.59) and (8.102)), but it is different in the Matausek–

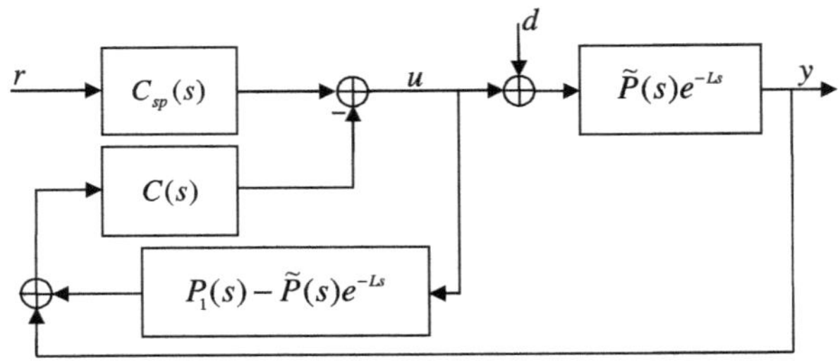

Fig. 8.33 Block diagram of the Chien–Peng–Liu modified Smith predictor

Micic scheme (see (8.74)), where the term K_r appears so that a full decoupling between the set-point response and the load disturbance response is not achieved. Further, the presence of the dead-time term in the denominator can yield a slow step disturbance response.

Summarising, the Normey-Rico–Camacho scheme appears to provide the same (high) performance of the Aström–Hang–Lim scheme by just simply modifying the scheme of Figure 8.4 by adding an appropriate set-point filter.

8.6 Chien–Peng–Liu Modified Smith Predictor

An approach similar to the method presented in Section 8.5, which is based on the use of a set-point filter, has been proposed in [16]. It consists of employing the scheme of Figure 8.4, with $P_1(s) = K(1-Ls)/s$, and of using a set-point weight β in the PI controller (see Section 2.1.2). The control scheme can be therefore modified as shown in Figure 8.33, where

$$C(s) = K_p\left(1+\frac{1}{T_i s}\right) \tag{8.110}$$

and

$$C_{sp}(s) = K_p\left(\beta+\frac{1}{T_i s}\right). \tag{8.111}$$

Thus, the set-point and load disturbance response is given by

$$\begin{aligned} Y(s) = {} & \frac{K_p\left(\beta+\frac{1}{T_i s}\right)\frac{K}{s}e^{-Ls}}{1+K_p\left(1+\frac{1}{T_i s}\right)\frac{K}{s}(-Ls+1)}R(s) \\ & + \frac{K}{s}e^{-Ls}\left(1-\frac{K_p\left(1+\frac{1}{T_i s}\right)\frac{K}{s}e^{-Ls}}{1+K_p\left(1+\frac{1}{T_i s}\right)\frac{K}{s}(-Ls+1)}\right)D(s). \end{aligned} \tag{8.112}$$

Focusing on the set-point response, Expression (8.112) can be simplified to

$$\frac{Y(s)}{R(s)} = \frac{KK_p(\beta T_i s+1)e^{-Ls}}{T_i(1-KK_pL)s^2+KK_p(T_i-L)s+KK_p}, \tag{8.113}$$

which is equal to

$$H(s) = \frac{1}{T_r s + 1} e^{-Ls} \tag{8.114}$$

by taking

$$K_p = \frac{1}{K(L + \beta T_r)} \tag{8.115}$$

and

$$T_i = \frac{T_r + L}{1 - \beta}. \tag{8.116}$$

The set-point weight β can then be employed to meet the control requirements in the load disturbance response. Indeed, by considering (8.112), (8.115), and (8.116), it can be derived that

$$\frac{Y(s)}{D(s)} = \frac{K}{s} e^{-Ls} \left(1 - \frac{\left(\left(\frac{T_r + L}{1-\beta} \right) s + 1 \right) e^{-Ls}}{\left(\frac{\beta T_r (T_r + L)}{1-\beta} \right) s^2 + \left(\frac{T_r + \beta L}{1-\beta} \right) s + 1} \right). \tag{8.117}$$

If the desired closed-loop response is selected as

$$\left(\frac{Y(s)}{D(s)} \right)_d = \frac{K}{s} e^{-Ls} \left(1 - \frac{((2T_l \xi + L) s + 1) e^{-Ls}}{T_l^2 s^2 + 2T_l \xi s + 1} \right), \tag{8.118}$$

the time constant T_l and the damping coefficient ξ of the second-order load response can be obtained from (8.117) by selecting

$$T_l = \sqrt{\frac{\beta T_r (T_r + L)}{1 - \beta}} \tag{8.119}$$

and

$$\xi = \frac{T_r + \beta L}{2 - \beta} \sqrt{\frac{1 - \beta}{\beta T_r (T_r + L)}}. \tag{8.120}$$

Thus, the tuning procedure consists in first selecting the closed-loop servo response speed by choosing appropriately the time constant T_r and then in choosing an appropriate value of β in order to obtain satisfactory values for the time constant T_l and the damping coefficient ξ (namely, in order to obtain a satisfactory load disturbance response).

It is worth stressing that the tuning procedure can be quite complex because of the nonlinear relations between T_l, ξ, and β. In case uncertainty in the dead time estimation is considered, a guideline can be obtained by applying a simple analysis based on the Routh–Hurwitz criterion [16]. The stability region for the parameter β results to be

$$1 > \beta > \frac{\Delta L}{T_r}, \tag{8.121}$$

where ΔL is the estimation error, and the usual assumption $T_r > \Delta L$ is made.

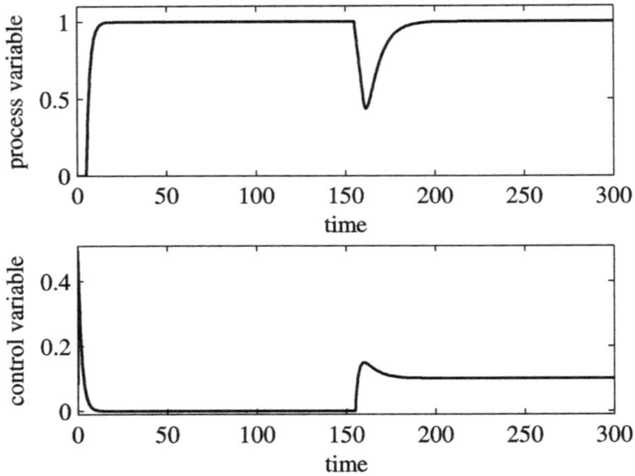

Fig. 8.34 Results obtained by applying the Chien–Peng–Liu modified Smith predictor (nominal case)

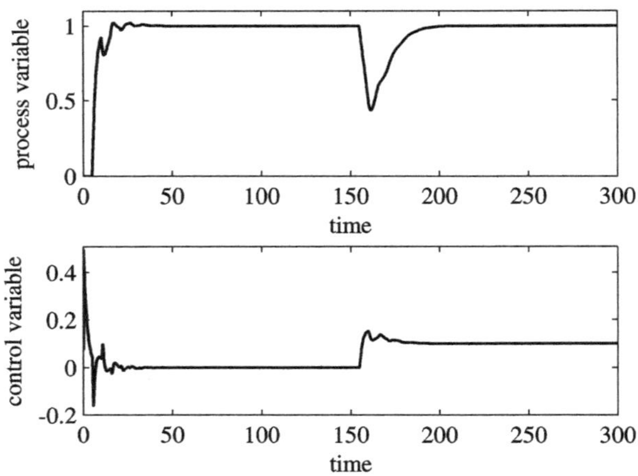

Fig. 8.35 Results obtained by applying the Chien–Peng–Liu modified Smith predictor (perturbed case)

The same process (8.5) is considered as an illustrative example. Then, $T_r = 2$ is chosen, and then β is selected equal to 0.5. As a consequence, $K_p = 0.17$, $T_i = 14$, and the expected values of T_l and ξ are 3.74 and 1.31, respectively. The set-point and load disturbance step responses are shown in Figure 8.34. In case the dead time is estimated as $L = 5.5$ (that is, $\Delta L = 0.5$), it results, with the same choice as before for T_r and β, $K_p = 0.15$, $T_i = 15$, and the expected values of T_l and ξ are 3.87 and 1.42, respectively. The corresponding simulation results are plotted in Figure 8.35.

8.7 Seshagiri Rao–Rao–Chidambaram Modified Smith Predictor

A generalisation of the Matausek–Micic modified Smith predictor, where the controller K_r is substituted by a PI controller $C(s)$ with set-point weight and $M(s)$ is a PD controller (see Figure 8.20) has been proposed in [106]. The tuning of the PI controller can be performed by direct synthesis. In fact, based on the transfer function between the set-point and the process output (see (8.74))

$$H(s) = \frac{Y(s)}{R(s)} = \frac{C(s)\tilde{P}(s)e^{-Ls}}{1 + C(s)\tilde{P}(s)}, \tag{8.122}$$

in which the dead time is outside of the feedback loop, a desired closed-loop transfer function can be specified as

$$\left(\frac{Y(s)}{R(s)}\right)_d = \frac{\eta s + 1}{(\lambda s + 1)^2}, \tag{8.123}$$

and therefore the controller transfer function can be determined as

$$C(s) = \frac{1}{\tilde{P}(s)} \frac{\left(\frac{Y(s)}{R(s)}\right)_d}{1 - \left(\frac{Y(s)}{R(s)}\right)_d} = \frac{\eta s + 1}{K(\lambda^2 s + (2\lambda - \eta))}. \tag{8.124}$$

By taking $\eta = 2\lambda$, Expression (8.124) can be reduced to a PI controller (8.85) with

$$K_p = \frac{2}{K\lambda}, \qquad T_i = 2\lambda. \tag{8.125}$$

The possibly high overshoot determined by the presence of the zero in the closed-loop transfer function (8.123) can be avoided by employing a set-point weight, which is suggested, based on a large number of simulations, to be selected in the range between 0.4 and 0.6.

The design of

$$M(s) = K_0\,(T_d s + 1) \tag{8.126}$$

can be performed based on gain and phase margin criteria. In particular, by considering that the characteristic equation involving $M(s)$ is

$$1 + M(s)P(s) = 0, \tag{8.127}$$

the open-loop transfer function is $M(s)P(s)$, and therefore, by the definitions of gain and phase margins, it can be written

$$\arg\big[M(j\omega_p)P(j\omega_p)\big] = -\pi, \tag{8.128}$$

$$\big|M(j\omega_g)P(j\omega_g)\big| = 1, \tag{8.129}$$

$$A_m = \frac{1}{|M(j\omega_p)P(j\omega_p)|}, \tag{8.130}$$

$$\phi_m = \pi + \arg\big[M(j\omega_g)P(j\omega_g)\big], \tag{8.131}$$

that is,

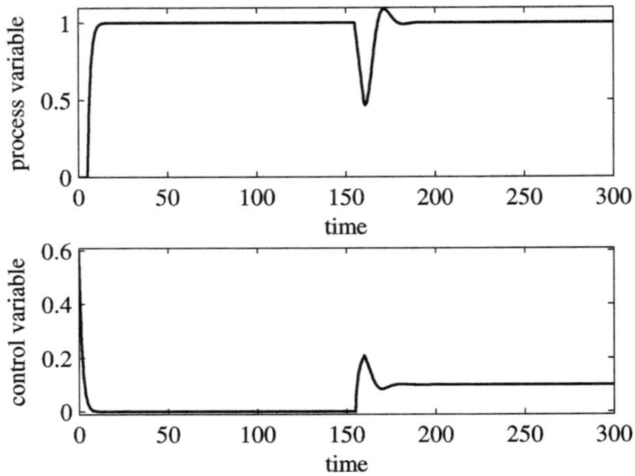

Fig. 8.36 Results obtained by applying the Seshagiri Rao–Rao–Chidambaram modified Smith predictor (nominal case)

$$\arctan(T_d\omega_p) - L\omega_p = -\frac{\pi}{2}, \tag{8.132}$$

$$\frac{K}{\omega_g}\sqrt{K_0\big(K_0 + T_d\omega_g^2\big)} = 1, \tag{8.133}$$

$$A_m = \frac{\omega_p}{K\sqrt{K_0\big(K_0 + T_d\omega_p^2\big)}}, \tag{8.134}$$

$$\phi_m = \frac{\pi}{2} + \arctan\big(T_d\omega_g\big) - L\omega_g, \tag{8.135}$$

where ω_g and ω_c are, evidently, the gain and phase crossover frequencies. Given desired values for A_m and ϕ_m, system (8.132)–(8.135) can be solved numerically for K_0 and T_d. In order to simplify the procedure, it is worth adopting the approximation

$$\arctan(x) = \frac{\pi}{2} - \frac{1}{x}. \tag{8.136}$$

It appears then that the overall tuning procedure is much more complex than that of Section 8.4. However, this method can be extended to processes with a double integrator.

The results obtained by applying the proposed controller to process (8.5) are shown in Figure 8.36, where $\lambda = 2$ has been chosen, and therefore (see (8.125)), the values $K_p = 1$ and $T_i = 4$ are obtained (the set-point weight has been fixed to 0.6). Further, as a consequence of the desired values $A_m = 1.8$ and $\phi_m = 55°$, the PD parameters are obtained as $K_0 = 0.152$ and $T_d = 1.493$. If the perturbed case is considered, namely, the estimated dead time is $L = 5.5$, the parameters obtained are $K_p = 1$, $T_i = 4$, $K_0 = 0.138$, and $T_d = 1.642$ (with the same desired gain and phase margin), and the corresponding set-point and load disturbance step responses are plotted in Figure 8.37.

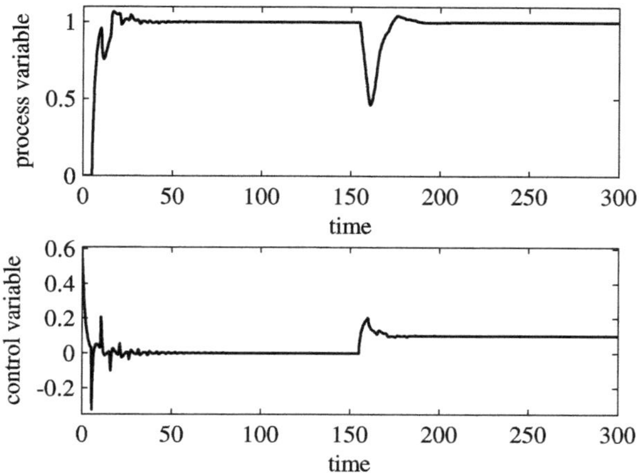

Fig. 8.37 Results obtained by applying the Seshagiri Rao–Rao–Chidambaram modified Smith predictor (perturbed case)

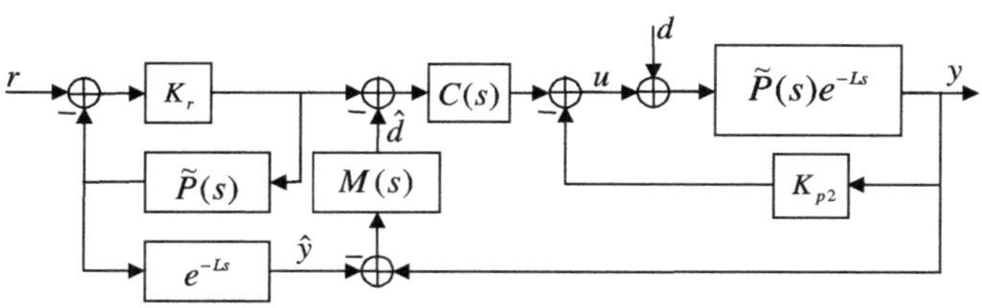

Fig. 8.38 Block diagram of the Tian–Gao modified Smith predictor

8.8 Tian–Gao Modified Smith Predictor

The method proposed by Tian and Gao in [120] consists of applying a modified Smith predictor scheme to a IPDT process that has been prestabilised by applying a local proportional feedback controller K_{p2}. The overall control scheme is shown in Figure 8.38, where $C(s)$ is a PI controller with a delayed integral action (see Section 7.4.2), and $M(s) = K_0\,(T_d s + 1)$ is a PD controller. The tuning of the proportional controller K_0 is performed first, by considering the inner loop transfer function

$$H_2(s) = \frac{\frac{K}{s}e^{-Ls}}{1 + \frac{K_{p2}K}{s}e^{-Ls}}. \tag{8.137}$$

Then, the same reasoning done in Section 8.4.1, based on the analysis of the characteristic equation, can be applied also in this case, yielding (see Expression (8.71))

$$K_{p2} = \frac{1}{2KL}. \tag{8.138}$$

Then, $C(s)$ is designed to be the denominator of the transfer function $H_2(s)$, that is,

$$C(s) = 1 + \frac{K_{p2}K}{s}e^{-Ls}. \tag{8.139}$$

In this way, the effect of K_{p2} in the set-point following task is eliminated. With the previous choices, the process output can be written as

$$\begin{aligned} Y(s) = &\frac{K_r K e^{-Ls}}{s + KK_r} R(s) \\ &+ \frac{Kse^{-Ls}}{(s + K_{p2}Ke^{-Ls})(s + KM(s)e^{-Ls})} D(s). \end{aligned} \tag{8.140}$$

Thus, the gain K_r can be tuned, by following the same reasoning of Section 8.4.1, as

$$K_r = \frac{1}{KT_r}. \tag{8.141}$$

The PD parameters K_0 and T_d can finally be selected by considering, similarly to Section 8.4.2, the characteristic equation

$$1 + \frac{KK_0(1 + T_d s)}{s}e^{-Ls} \tag{8.142}$$

and by considering the corresponding phase margin relations:

$$\frac{KK_0\sqrt{1 + T_d^2\omega_c^2}}{\omega_c} = 1, \tag{8.143}$$

$$\phi_m = \frac{\pi}{2} + \arctan(T_d\omega) - \omega L. \tag{8.144}$$

Then, by selecting

$$T_d = \alpha L, \tag{8.145}$$

Equations (8.143) and (8.144) can be solved for K_0 and T_d after having chosen for α and ϕ_m the suggested values of 0.5 and 60°, respectively.

The results related to the control of process (8.5) in the nominal case are shown in Figure 8.39. Note that the application of the tuning procedure presented above yields $K_{p2} = 1$, $K_r = 0.5$ (T_r has been fixed to 2 as in the other examples of this chapter), $K_0 = 0.176$, and $T_d = 2.5$. Conversely, if the dead time is estimated as $L = 5.5$, the design parameters are modified as $K_{p2} = 0.09$, $K_0 = 0.160$, and $T_d = 2.75$. The corresponding set-point and load disturbance step responses are plotted in Figure 8.40.

8.9 More Complex Schemes

Other more complex schemes have been proposed in the last years. A few examples of them are presented hereafter.

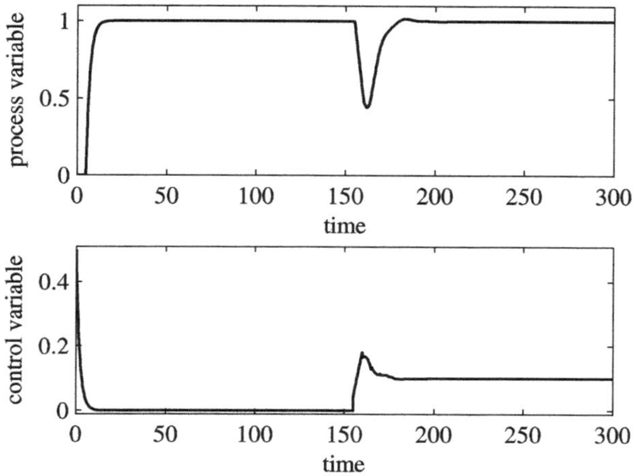

Fig. 8.39 Results obtained by applying the Tian–Gao modified Smith predictor (nominal case)

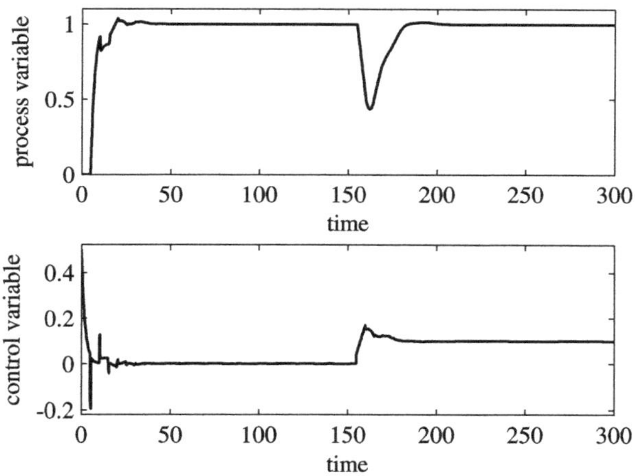

Fig. 8.40 Results obtained by applying the Tian–Gao modified Smith predictor (perturbed case)

8.9.1 Majhi–Atherton Modified Smith Predictor

In the control scheme proposed in [62–65], which is shown in Figure 8.41, a new proportional controller K_{p2} is added to the control scheme of Figure 8.20 with the aim of stabilising the integral process, and $C(s)$ is a PI controller (see (8.85)).

The resulting set-point and load disturbance response is given as (by always assuming that the process is described by an IPDT model (8.1))

$$Y(s) = \frac{C(s)P(s)}{1+\tilde{P}(s)(C(s)+K_{p2})}R(s)$$

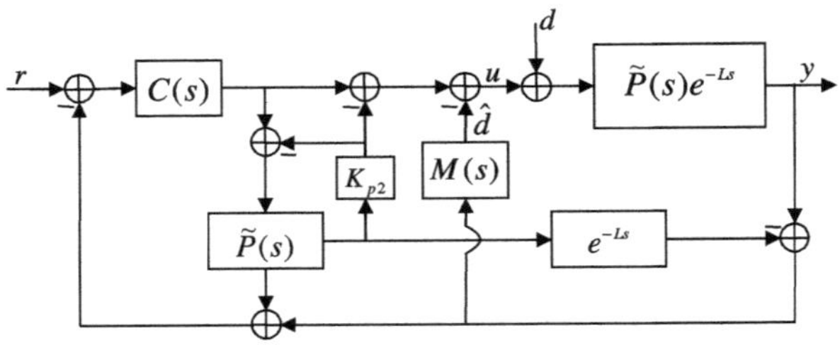

Fig. 8.41 Block diagram of the Tian–Gao modified Smith predictor

$$+\frac{P(s)}{1+\tilde{P}(s)(C(s)+K_{p2})}\frac{1+\tilde{P}(s)(C(s)+K_{p2})-C(s)P(s)}{1+M(s)P(s)}D(s). \tag{8.146}$$

The method described in [65] can be used for the design of the three controllers employed in the control scheme. By first constraining the value of K_{p2} as

$$K_{p2}=\frac{1}{K}, \tag{8.147}$$

the tuning of the PI controller $C(s)$ can be performed by first setting

$$T_i=\frac{1}{KK_{p2}}=1. \tag{8.148}$$

In this way, the transfer function between the set-point and the process output is

$$H(s)=\frac{Y(s)}{R(s)}=\frac{1}{\frac{1}{KK_p}s+1}e^{-Ls}, \tag{8.149}$$

and therefore the proportional gain K_p can be selected appropriately in order to obtain the desired closed-loop time constant as in Section 8.4.1.

Finally, the transfer function $M(s)$ is chosen as a simple proportional gain, $M(s)=K_0$, whose value can be determined by following the same passages of Section 8.4.1 that yield

$$K_0=\frac{\frac{\pi}{2}-\phi_m}{KL}. \tag{8.150}$$

In [65] it is suggested to select a phase margin $\phi_m=60°$, namely,

$$K_0=\frac{\pi}{6KL}=\frac{0.5236}{KL}. \tag{8.151}$$

The usual process (8.5) is considered as a worked example. By selecting a desired closed-loop time constant equal to 2, the resulting value of the proportional gain is $K_p=0.5$, and by applying the tuning rules (8.147)–(8.148) and (8.151) the resulting values of the other controller parameters are $K_{p2}=1$, $T_i=1$, and $K_0=0.105$. The set-point unit step response and the response to a load step disturbance of amplitude –0.1 applied at $t=150$ are plotted in Figure 8.42. If the dead time is estimated

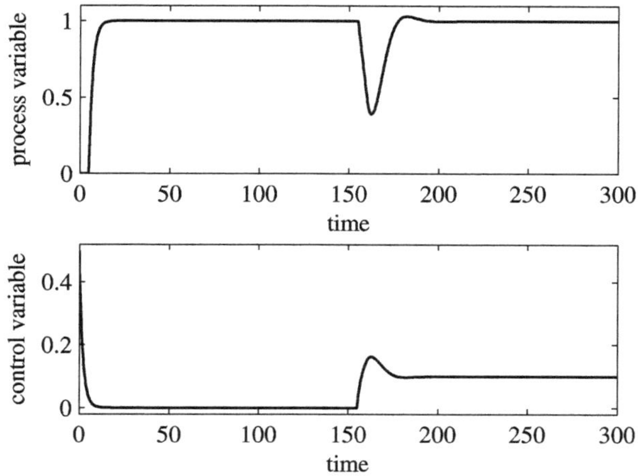

Fig. 8.42 Results obtained by applying the Majhi–Atherton modified Smith predictor (nominal case)

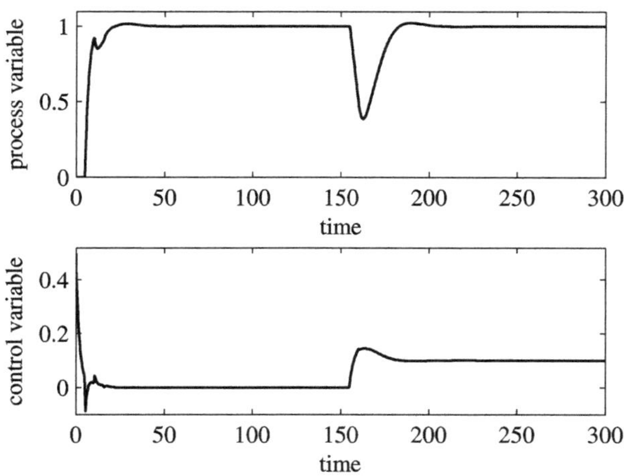

Fig. 8.43 Results obtained by applying the Majhi–Atherton modified Smith predictor (perturbed case)

as $L = 5.5$, the only controller parameter that is modified is $K_0 = 0.095$, and the corresponding results are in Figure 8.43.

It is worth stressing at this point that the presented design method can be applied also in an automatic tuning context [64, 65] and that a tuning method based on standard forms has also been proposed [62, 63].

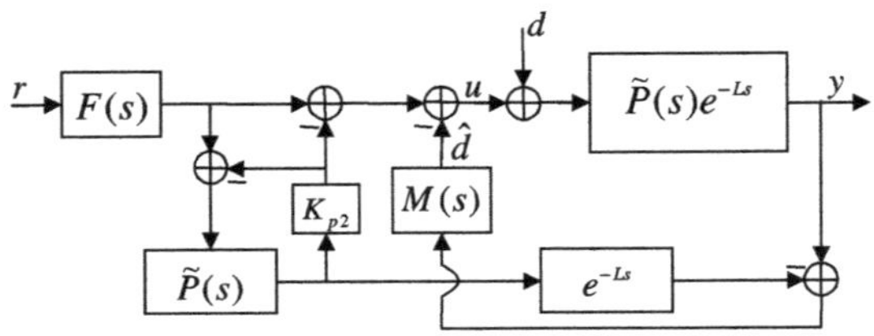

Fig. 8.44 Block diagram of the Liu–Cai–Gu–Zhang modified Smith predictor

8.9.2 Liu–Cai–Gu–Zhang Modified Smith Predictor

In the modified Smith predictor of Section 8.9.1 the load disturbance transfer function is somewhat complicated, and the controllers $C(s)$ and K_{p2} appear in both the set-point and the load disturbance transfer function, and therefore the two control tasks are coupled from the design point of view. In order to simplify the design and to decouple the two control tasks, a solution has been proposed in [60], which consists of using a control scheme where, with respect to that of Figure 8.41, the feedback signals from the actual process output and the model output to the set-point are cut off. The new scheme is shown in Figure 8.44.

The proportional gain K_{p2} can be tuned by considering the transfer function between the set-point and the process output (in the nominal case)

$$H(s) = \frac{Y(s)}{R(s)} = \frac{F(s)P(s)}{1 + K_{p2}\tilde{P}(s)}. \tag{8.152}$$

The characteristic equation is therefore

$$s + KK_{p2} = 0, \tag{8.153}$$

which implies that the closed-loop system is stable for $K_{p2} > 0$. A simple choice is therefore to use a unity-feedback between the input and output of the delay-free part of the process model, that is, to use

$$K_{p2} = 1. \tag{8.154}$$

The prefilter $F(s)$ can then be designed, as in Section 8.5.4, in order to minimise the H_2 norm of the error signal, that is, in order to solve the following optimisation problem:

$$\min \left\| \Gamma(s)\big(1 - H(s)\big) \right\|_2^2, \tag{8.155}$$

where $\Gamma(s) = 1/s$ (see Section 2.3.4.2). It results in

$$F(s) = \frac{s + KK_{p2}}{K\,(\lambda s + 1)}, \tag{8.156}$$

which is optimal for $\lambda \to 0$. In practice, λ is the desired closed-loop time constant.

The disturbance estimator $M(s)$ can be designed by first considering that the load disturbance transfer function between the load disturbance d and its estimation $\hat{d}$ is

$$W(s) = \frac{\hat{d}}{d} = \frac{M(s)P(s)}{1 + M(s)P(s)}. \tag{8.157}$$

Ideally, this transfer function should be equal to e^{-Ls}, that is, when a load disturbance occurs, the controller should produce an inversely equivalent signal to offset it just after the process dead time L. However, by taking into account the steady-state requirements, the desired transfer function can be selected as

$$W_d(s) = \frac{as+1}{(\lambda_d s+1)^2} e^{-Ls}, \tag{8.158}$$

and, in order to obtain a zero steady-state error, it has to be (the sensitivity function should have two zeros at $s = 0$)

$$\lim_{s \to 0} \frac{d}{ds}\left[1 - W_d(s)\right] = 0, \tag{8.159}$$

that is,

$$a = 2\lambda_d + L. \tag{8.160}$$

From Expression (8.157) the corresponding transfer function $M(s)$ can be determined as

$$M(s) = \frac{W_d(s)}{1 - W_d(s)} \frac{1}{P(s)} = \frac{s(as+1)}{K((\lambda_d s+1)^2 - (as+1)\,e^{-Ls})}. \tag{8.161}$$

However, it can be noted that a pole-zero cancellation occurs at $s = 0$. In order to avoid this internal instability and in order to obtain a more practical transfer function, the Maclaurin expansion can be employed to rewrite the estimator transfer function. Thus, $M(s)$ can be rewritten as

$$M(s) = \frac{\overline{M}(s)}{s}, \tag{8.162}$$

and, therefore,

$$M(s) = \frac{1}{s}\left[\overline{M}(0) + \overline{M}'(0)s + \frac{\overline{M}''(0)}{2!}s^2 + \cdots\right]. \tag{8.163}$$

It can be easily noted that, by taking into account the first three terms only, a PID controller results, namely, $M(s)$ can be written as

$$M(s) = K_{pM} + \frac{K_{iM}}{s} + K_{dM}s, \tag{8.164}$$

where

$$K_{pM} = \overline{M}'(0), \qquad K_{iM} = \overline{M}(0), \qquad K_{dM} = \frac{\overline{M}''(0)}{2}. \tag{8.165}$$

The tuning of the disturbance estimator depends only on the parameter λ_d, which can be conveniently selected based on the desired trade-off between aggressiveness and robustness.

In the worked example with process (8.5), after having selected $\lambda = 2$ and $\lambda_d = 3$ (and $K_{p2} = 1$), the load estimator transfer function results to be (note that a high-frequency low-pass filter has to be added to make it proper)

$$M(s) = 0.2576 + \frac{1}{51.5s} + 0.5069s. \tag{8.166}$$

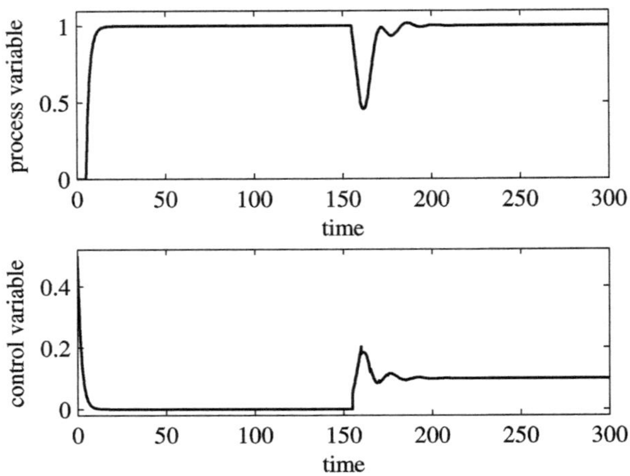

Fig. 8.45 Results obtained by applying the Liu–Cai–Gu–Zhang modified Smith predictor (nominal case)

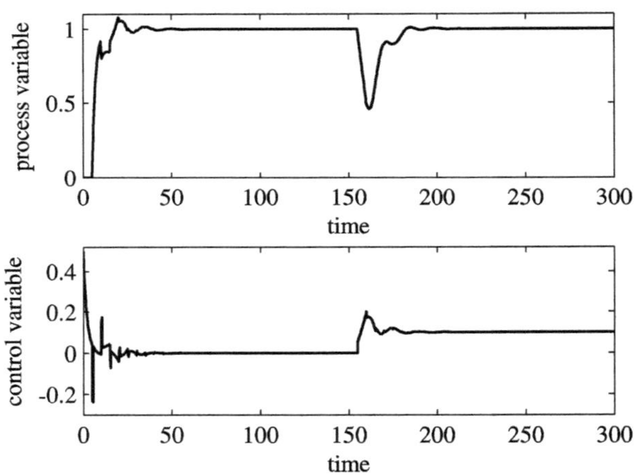

Fig. 8.46 Results obtained by applying the Liu–Cai–Gu–Zhang modified Smith predictor (perturbed case)

The results obtained in the set-point and load disturbance step response are plotted in Figure 8.45. If the dead time is estimated as $L = 5.5$, it results

$$M(s) = 0.2461 + \frac{1}{57.1s} + 0.5439s, \tag{8.167}$$

and the corresponding set-point and load disturbance step responses are shown in Figure 8.46.

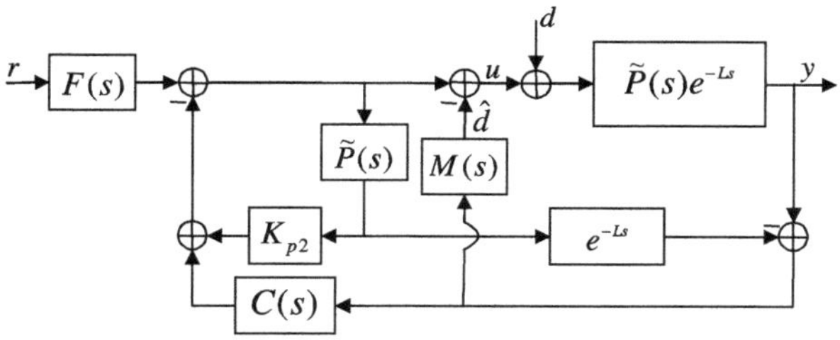

Fig. 8.47 Block diagram of the Lu–Yang–Wang–Zheng modified Smith predictor

8.9.3 Lu–Yang–Wang–Zheng Modified Smith Predictor

A double two-degree-of-freedom modified Smith predictor has been proposed in [61]. The control scheme is shown in Figure 8.47, where it appears that there are four controllers. In particular, as in Sections 8.9.1 and 8.9.2, the proportional controller K_{p2} is employed to stabilise the delay-free process $\tilde{P}(s)$, and $M(s)$ is employed to estimate the load disturbance d. Then, the two additional controllers $F(s)$ and $C(s)$ are used to enhance the performance in the set-point following and load disturbance rejection task, respectively.

As in the previous sections, the design of the controller is made by considering that a perfect model of the process (8.1) is available. In this case, the transfer function between the set-point and the process output is

$$H(s) = \frac{F(s)\tilde{P}(s)}{1+K_{p2}\tilde{P}(s)}e^{-Ls}. \tag{8.168}$$

A desired closed-loop transfer function

$$\frac{1}{T_r s+1}e^{-Ls} \tag{8.169}$$

can then be obtained easily by setting

$$K_{p2} = \frac{1}{KT_r}, \qquad F(s) = \frac{1}{KT_r}, \tag{8.170}$$

that is, $F(s)$ is a simple proportional controller. With the previous choices, the load disturbance closed-loop transfer function is

$$H_d(s) = \frac{KT_r\left(s+\frac{1}{T_r}-KC(s)e^{-Ls}\right)}{(T_r s+1)\,(s+KM(s)e^{-Ls})}e^{-Ls}. \tag{8.171}$$

By selecting $M(s) = K_0$ as a simple proportional controller, its value can be determined as in (8.151):

$$K_0 = \frac{\pi}{6KL} = \frac{0.5236}{KL}. \tag{8.172}$$

Finally, if $C(s)$ is chosen as a PD controller, namely,

$$C(s) = K_p\,(1+T_d s)\,, \tag{8.173}$$

then Expression (8.171) can be rewritten as

$$H_d(s) = \frac{KT_r\left(s + \frac{1}{T_r} - KK_p(1+T_d s)\,e^{-Ls}\right)}{(T_r s+1)\,(s+KK_0 e^{-Ls})} e^{-Ls}. \tag{8.174}$$

By employing the approximation $e^{-Ls} \cong 1 - Ls$ we have

$$H_d(s) = \frac{KT_r\left(KLK_pT_d s^2 + (-KK_pT_d + KK_pL + 1)s + \left(\frac{1}{T_r} - KK_p\right)\right)}{T_r\left(1-\frac{\pi}{6}\right)s^2 + \left(T_r\frac{\pi}{6L} + 1 - \frac{\pi}{6}\right)s + \frac{\pi}{6L}} e^{-Ls}. \tag{8.175}$$

In order to have a null steady-state error in the presence of a constant load disturbance, the constant term in the numerator has to be zero, so that

$$K_p = \frac{1}{KT_r}. \tag{8.176}$$

The load disturbance unit step response can be therefore expressed as

$$Y_d(s) = \frac{1}{s}H_d(s) = KT_r\frac{KLK_pT_d s + (-KK_pT_d + KK_pL + 1)}{T_r\left(1-\frac{\pi}{6}\right)s^2 + \left(T_r\frac{\pi}{6L} + 1 - \frac{\pi}{6}\right)s + \frac{\pi}{6L}} e^{-Ls}, \tag{8.177}$$

and in this context, T_d can be chosen in order to minimise the integrated square error, that is [61],

$$T_d = \frac{T_r\,(6-\pi)\,(T_r+L)}{T_r\,(6-\pi) + L\pi}. \tag{8.178}$$

Alternatively, if the first-order Padè approximation $e^{-Ls} \cong (1 - Ls/2)/(1 + Ls/2)$, by applying the same passages, the value of K_p is still determined as (8.176), while the derivative time constant results to be

$$T_d = T_r\frac{\left(\frac{L}{2} + T_r - \frac{\pi L}{12}\right)\left(1 + \frac{L}{T_r}\right) - \frac{\pi}{24}}{\frac{L}{2} + T_r - \frac{T_r\pi}{12} + \frac{\pi L}{24}}. \tag{8.179}$$

If the control system is applied to process (8.5), the proposed method yields (when $T_r = 2$, as in the previous cases) $K_{p2} = F(s) = 0.5$ and $K_0 = 0.105$. Then, $C(s) = 0.5\,(1 + 1.868s)$ if Expression (8.178) is employed, and $C(s) = 0.5\,(1 + 5.728s)$ if Expression (8.179) is employed. The resulting set-point and load disturbance step response in the two cases is plotted in Figure 8.48. The performance achieved is satisfactory in the nominal case, but if the dead time is estimated as $L = 5.5$, the control system becomes unstable, which implies that it is very sensible to modelling uncertainties.

8.10 Conclusions

In this chapter a large number of modified Smith predictor schemes have been presented. The need of modifying the standard Smith predictor comes from the fact that the standard Smith predictor is not capable, for an integral process, of providing a

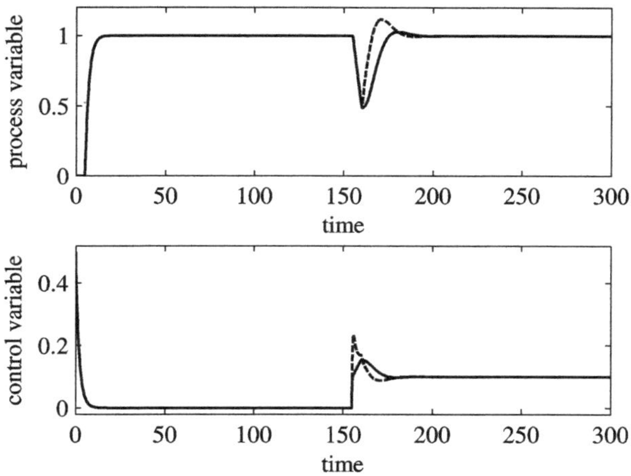

Fig. 8.48 Results obtained by applying the Lu–Yang–Wang–Zheng modified Smith predictor (nominal case). *Solid line*: PD controller given by Expression (8.178). *Dashed line*: PD controller given by Expression (8.179)

null steady-state error in the presence of a constant load disturbance. The different rationale in the design of the modified Smith predictors, as well as their different levels of complexity, have been highlighted. The presented illustrative examples have shown the effectiveness of the proposed control architectures and that there is no big difference in the achieved performance.

Chapter 9
Smith-principle-based PID-type Control

It has been shown in the previous chapter that Smith-predictor-based schemes are very effective for processes with dead-time. On the other hand, PID controllers are the most widely used controllers in industry [4, 5]. In this chapter, a control scheme which combines their advantages as proposed in [161] is described. The controller is inherently a PID-type controller in which the integral action is implemented with a delay unit rather than a pure integrator while retaining the advantage of the Smith predictor, *i.e.*, the Smith principle. The set-point response and the disturbance response are decoupled from each other and can be designed separately. Another advantage of this control scheme is that the robustness is easy to analyse and can be guaranteed explicitly, compromising the robustness with the disturbance response.

9.1 The Control Scheme

The control scheme is shown in Figure 9.1. It consists of a prefilter $F(s)$, a main controller $C(s)$, and a delay unit cascaded with a low-pass filter $Q(s)$. The prefilter $F(s)$ is the first degree-of-freedom, and the low-pass filter $Q(s)$ is the second degree-of-freedom.

The local positive feedback loop with $Q(s)e^{-Ls}$ has been widely used in a repetitive control scheme [33, 147] to control processes *without* dead time. Repetitive control is a technique using delay elements to improve some system performance, in particular, the tracking accuracy and/or the disturbance-rejection of periodic signals. With the learning ability of delay elements, repetitive control can reject/track any periodic signals of period L. Hence, to some extent, the control scheme under consideration is a kind of repetitive control for processes *with* dead time. The delay element used here is equal to the dead time that exists in the process rather than the period of a periodic signal to be tracked or rejected. Obviously, a by-product of this control scheme is that if the bandwidth of $Q(s)$ is wide enough, it can track/reject periodic signals with the period equal to the dead time.

A. Visioli, Q.-C. Zhong, *Control of Integral Processes with Dead Time*, Advances in Industrial Control,
DOI 10.1007/978-0-85729-070-0_9, © Springer-Verlag London Limited 2011

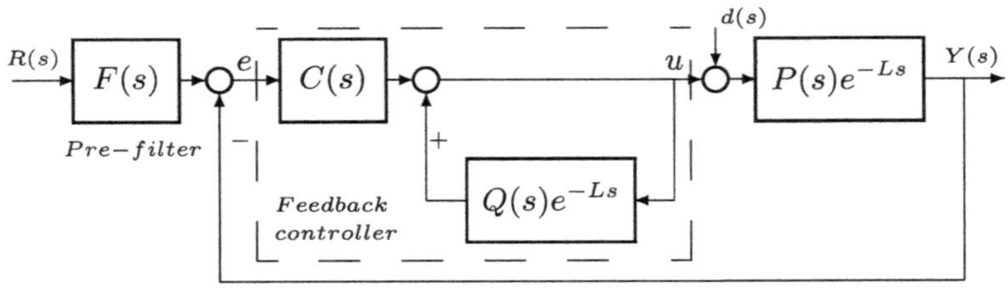

Fig. 9.1 A two-degree-of-freedom controller for processes with dead time

The transfer function from the set-point $R(s)$ to the output $Y(s)$ is

$$G_r(s) = F(s)\frac{C(s)\frac{1}{1-Q(s)e^{-Ls}}P(s)e^{-Ls}}{1+C(s)\frac{1}{1-Q(s)e^{-Ls}}P(s)e^{-Ls}}$$
$$= F(s)\frac{C(s)P(s)e^{-Ls}}{1-Q(s)e^{-Ls}+C(s)P(s)e^{-Ls}}. \tag{9.1}$$

If the main controller $C(s)$ is designed as

$$C(s) = Q(s)P^{-1}(s), \tag{9.2}$$

then, in the nominal case,

$$G_r(s) = F(s)Q(s)e^{-Ls}. \tag{9.3}$$

If the prefilter $F(s)$ is designed as

$$F(s) = \frac{1}{\lambda s+1}Q^{-1}(s) \tag{9.4}$$

or

$$F(s) = \frac{1}{\lambda^2 s^2 + 2\lambda\zeta s + 1}Q^{-1}(s), \tag{9.5}$$

then the desired set-point response is obtained as

$$G_r(s) = \frac{1}{\lambda s+1}e^{-Ls} \quad \text{or} \quad G_r(s) = \frac{1}{\lambda^2 s^2 + 2\lambda\zeta s + 1}e^{-Ls} \quad (\lambda > 0, \zeta > 0), \tag{9.6}$$

which is independent of the second degree-of-freedom $Q(s)$. The disturbance response of the system is

$$G_d(s) = \frac{P(s)e^{-Ls}}{1+C(s)\frac{1}{1-Q(s)e^{-Ls}}P(s)e^{-Ls}}$$
$$= \left(1-Q(s)e^{-Ls}\right)P(s)e^{-Ls}, \tag{9.7}$$

Table 9.1 Controllers for different desired responses ($\alpha > 0$)

$Q(s)$	$C(s)$	$F(s)$	Desired Response
$\frac{(2\alpha+L)s+1}{(\alpha s+1)^2}$	$\frac{(2\alpha+L)s^2+s}{K(\alpha s+1)^2}$	$\frac{(\alpha s+1)^2}{(\lambda s+1)((2\alpha+L)s+1)}$	$\frac{1}{\lambda s+1}e^{-Ls}$
		$\frac{(\alpha s+1)^2}{(\lambda^2 s^2+2\lambda\zeta s+1)((2\alpha+L)s+1)}$	$\frac{1}{\lambda^2 s^2+2\lambda\zeta s+1}e^{-Ls}$

which is independent of the first degree-of-freedom $F(s)$. Therefore, the set-point response and the disturbance response are decoupled from each other and can be designed separately.

In order to obtain zero static error under step set-point/disturbance change, $Q(s)$ should satisfy

$$Q(0) = 1 \tag{9.8}$$

for processes without an integrator and, furthermore,

$$\dot{Q}(0) = L \tag{9.9}$$

for processes with an integrator. In order to implement the controller physically, the relative degree of $Q(s)$ should not be less than that of $P(s)$, and the relative degree of the delay-free part of the desired response should not be less than that of $Q(s)$. Hence, in general, $Q(s)$ is a low-pass filter with unity static gain. In order to guarantee the stability of $Q^{-1}(s)$ and the internal stability of the closed-loop system, $Q(s)$ should be of minimum phase.

For typical IPDT process $P(s) = \frac{K}{s}e^{-Ls}$, the minimum order of $Q(s)$ should be 2 because of the constraint (9.9). Let

$$Q(s) = \frac{\beta s + 1}{(\alpha s + 1)^2}. \tag{9.10}$$

Then, according to (9.9),

$$\beta = 2\alpha + L. \tag{9.11}$$

The corresponding controllers needed to obtain a desired first-order or second-order response plus dead time (9.6) are given in Table 9.1.

Since there exists pole-zero cancellation at $s = 0$ between $C(s)$ and the plant, the structure shown in Figure 9.1 is not internally stable, and a structure which is internally stable needs to be found to implement the design.

9.2 An Equivalent Structure for Implementation

According to Table 9.1 and Figure 9.1, the feedback controller is

$$C(s)\frac{1}{1 - Q(s)e^{-Ls}} = \frac{(2\alpha + L)s^2 + s}{K(\alpha s + 1)^2}\frac{1}{1 - \frac{(2\alpha+L)s+1}{(\alpha s+1)^2}e^{-Ls}}$$

$$\begin{aligned}
&= \frac{1}{K}\frac{(2\alpha+L)s^2+s}{(\alpha s+1)^2-(2\alpha+L)se^{-Ls}-e^{-Ls}} \\
&= \frac{1}{K}\frac{(2\alpha+L)s+1}{\alpha^2 s-L+(2\alpha+L)(1-e^{-Ls})+\frac{1-e^{-Ls}}{s}} \\
&= \frac{1}{K}\frac{(2\alpha+L)s+1}{\alpha^2 s-L+((2\alpha+L)s+1)\frac{1-e^{-Ls}}{s}} \\
&\approx \frac{1}{K}\frac{(2\alpha+L)s+1}{\alpha^2 s+(2\alpha+L)Ls} \\
&= \frac{(2\alpha+L)s+1}{K(\alpha+L)^2 s},
\end{aligned} \tag{9.12}$$

where the approximation $\frac{1-e^{-Ls}}{s} \approx L$ is used. Hence, the feedback controller is inherently a PI controller $K_p + \frac{K_i}{s}$ with

$$\begin{cases} K_p = \dfrac{2\alpha+L}{K(\alpha+L)^2}, \\ K_i = \dfrac{1}{K(\alpha+L)^2}. \end{cases} \tag{9.13}$$

The controller retains the advantages of the Smith predictor and the advantages of the PID controller. A prominent advantage of the tuning formula (9.13) is that there exists a free parameter α. This free parameter can be used to compromise the disturbance response with the robustness (as shown later in Section 9.4); it can also be used to optimise a certain performance index, *e.g.*, the gain or phase margin.

Moreover, according to (9.12), the feedback controller for the IPDT process can be rewritten as

$$C(s)\frac{1}{1-Q(s)e^{-Ls}} = \frac{1}{K}\frac{\frac{(2\alpha+L)s+1}{\alpha^2 s-L}}{1+\frac{(2\alpha+L)s+1}{\alpha^2 s-L}\frac{1-e^{-Ls}}{s}}. \tag{9.14}$$

This is a negative feedback loop of $\frac{(2\alpha+L)s+1}{\alpha^2 s-L}$ through a finite impulse response (FIR) block $\frac{1-e^{-Ls}}{s}$ cascaded by a gain $\frac{1}{K}$, as shown in Figure 9.2. The unstable pole-zero cancellation is avoided as long as the FIR block $\frac{1-e^{-Ls}}{s}$ is implemented non-dynamically [89, 158–160]; see Section 12.3. This feedback controller itself is stable if $\alpha > 0.63L$ [72].

9.3 Robustness Analysis

In the nominal case, the loop transfer function of the control system shown in Figure 9.1 is

$$W(s) = \frac{Q(s)e^{-Ls}}{1-Q(s)e^{-Ls}}. \tag{9.15}$$

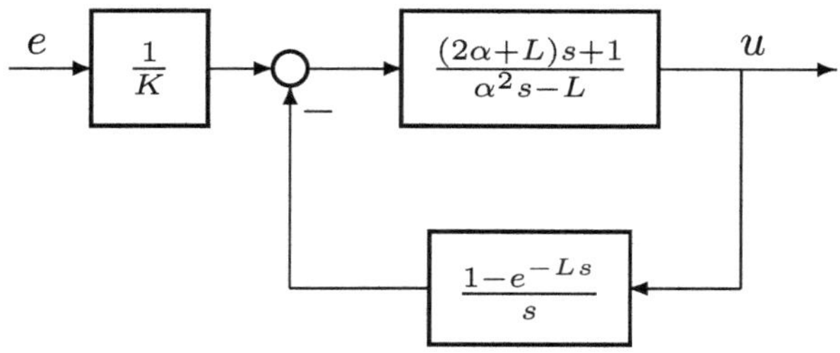

Fig. 9.2 Equivalent feedback controller for an IPDT process

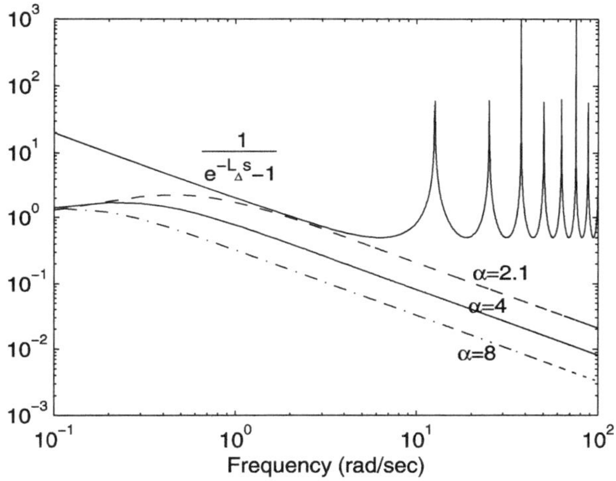

Fig. 9.3 Design of $Q(s)$

Hence, the corresponding sensitivity function is

$$S(s) = \frac{1}{1+W(s)} = 1 - Q(s)e^{-Ls}, \tag{9.16}$$

and the complementary sensitivity function is

$$T(s) = 1 - S(s) = Q(s)e^{-Ls}. \tag{9.17}$$

This means that $Q(s)$ does not only affect the disturbance response (9.7) but also affects the robust stability and the robust performances. It should be determined by compromising the disturbance response with the robustness.

Theorem 9.1 *Assume that there exists a multiplicative uncertainty $\Delta(s) \in H_\infty$ in the delay-free part. Then the closed-loop system is robustly stable if $\|Q(s)\|_\infty < \frac{1}{\|\Delta(s)\|_\infty}$.*

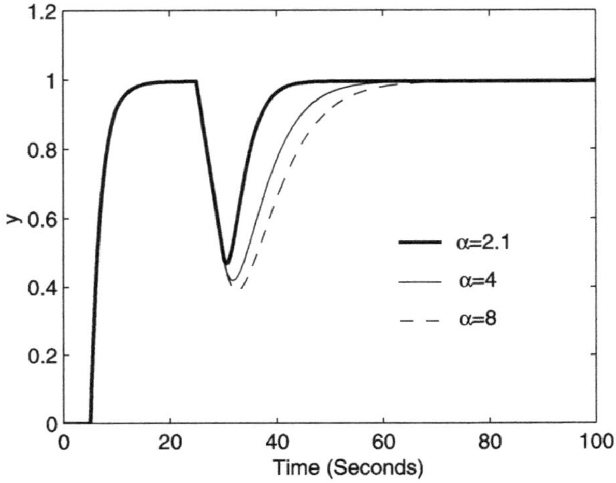

(a) Responses when $L_\Delta = 0$

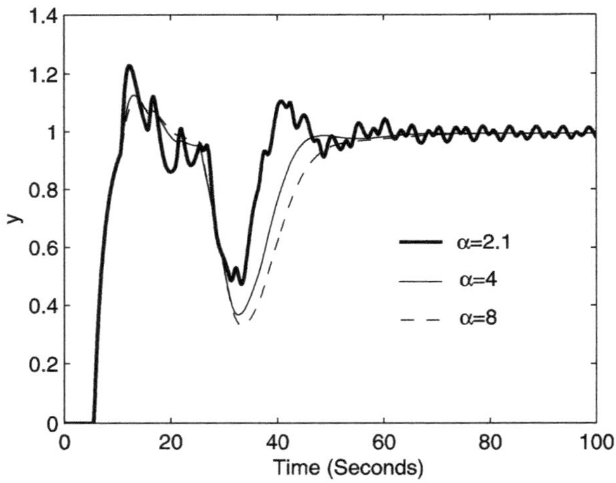

(b) Responses when $L_\Delta = 0.5$ seconds

Fig. 9.4 System responses

Proof Since

$$\begin{aligned}\|T(s)\Delta(s)\|_\infty &= \|Q(s)e^{-Ls}\Delta(s)\|_\infty \\ &= \|Q(s)\Delta(s)\|_\infty \\ &< \|Q(s)\|_\infty \cdot \|\Delta(s)\|_\infty \\ &< 1,\end{aligned}$$

the closed-loop system is internally stable for all $\Delta(s) \in H_\infty$ according to Theorem 8.5 in [165]. □

If there exists an uncertainty L_Δ in the dead time, then

$$\begin{aligned} G(s) &= P(s)e^{-(L+L_\Delta)s} \\ &\doteq P(s)\big(1+\Delta(s)\big)e^{-Ls} \end{aligned} \tag{9.18}$$

with

$$\Delta(s) = e^{-L_\Delta s} - 1. \tag{9.19}$$

Hence, the following results hold:

Corollary 9.1 *Assume that there exists an uncertainty $L_\Delta > -L$ in the dead-time. Then the closed-loop system is robustly stable if $\|Q(s)\|_\infty < \frac{1}{\|e^{-L_\Delta s}-1\|_\infty}$.*

Corollary 9.2 *Assume that there exist an uncertainty $L_\Delta > -L$ in the dead-time and a multiplicative uncertainty $\Delta(s) \in H_\infty$ in the delay-free part. Then the closed-loop system is robustly stable if $\|Q(s)\|_\infty < \frac{1}{\|(1+\Delta(s))e^{-L_\Delta s}-1\|_\infty}$.*

9.4 Simulation Examples

Consider

$$P(s) = \frac{1}{s}e^{-5s} \tag{9.20}$$

and assume that there exists a dead-time uncertainty $0 \le L_\Delta \le 0.5$ seconds.

The amplitude responses of the system for $Q(s) = \frac{(2\alpha+5)s+1}{(\alpha s+1)^2}$ with $\alpha = 2.1$, $\alpha = 4$, and $\alpha = 8$ are shown in Figure 9.3 together with that of the $\frac{1}{e^{-L_\Delta s}-1}$. The desired set-point response is designed with $\lambda = 2$. The responses in the nominal case are shown in Figure 9.4(a), and the responses when $L_\Delta = 0.5$ seconds are shown in Figure 9.4(b). In all cases, a step disturbance $d(t) = 0.1$ was applied at $t = 15$ seconds. The smaller the α, the better the disturbance response but the worse the robust performance. The larger the α, the better the robust performance but the worse the disturbance response.

9.5 Conclusions

A control scheme combining the advantages of a PID controller and a Smith predictor is presented in this chapter. It is a two-degree-of-freedom structure with the ability to decouple the set-point response and disturbance response from each other.

The dead time is eliminated from the characteristic equation of the closed-loop system, and only two or three parameters (in addition to model parameters) need to be tuned; either one or two of them (belonging to the degree-of-freedom $F(s)$) are determined by the desired response, and another one (belonging to the degree-of-freedom $Q(s)$) is determined by compromising the robustness with the disturbance response.

Chapter 10
Disturbance Observer-based Control

In this chapter, it is revealed that a disturbance observer-based control scheme is very effective in controlling integral processes with dead time. The controller can be designed to reject ramp disturbances, step disturbances, and even arbitrary disturbances. Only two parameters are left to tune when the plant model is available. One is the time constant of the set-point response, and the other is the time constant of the disturbance response. The latter is tuned to compromise the disturbance response with robustness. This control scheme has a simple, clear, easy-to-design, and easy-to-implement structure.

10.1 Disturbance Observer

A disturbance observer uses the inverse of the nominal plant model to observe the disturbance applied to the plant, which is then directly used to cancel the effect of the disturbance in the control signal. As a result, the closed loop system is forced to act as the nominal plant. It was originally presented by Ohnishi *et al.* [85] to handle disturbances in motion control. Umeno and Hori [123] refined it and applied it to the robust control of DC servo motors. Endo *et al.* [24] and Kempf and Kobayashi [51] applied it to control the high-speed direct-drive positioning table. In these papers, the dead time in the process was not included. Hong and Nam [36] explicitly considered the measurement delay in the load torque observer to improve the stability. The disturbance observer is then extended to IPDT processes in [156, 162, 163]. It is a version of the 2DOF internal model control [76], and it is able to decouple the disturbance response from the set-point response. The robust stability of the closed-loop system is quite easy to be guaranteed graphically.

A. Visioli, Q.-C. Zhong, *Control of Integral Processes with Dead Time*, Advances in Industrial Control,
DOI 10.1007/978-0-85729-070-0_10, © Springer-Verlag London Limited 2011

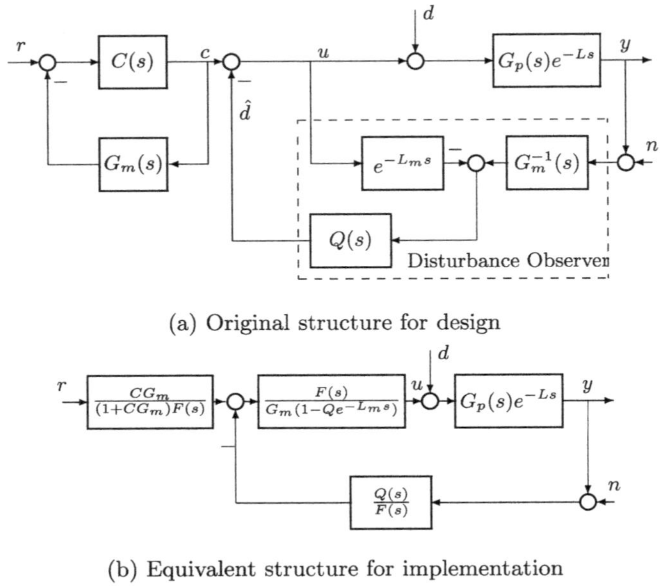

Fig. 10.1 Disturbance observer-based control scheme

10.2 Control Structure

Consider the following integral process with dead time:

$$G(s) = G_p(s)e^{-Ls} = \frac{G_{p0}(s)}{s}e^{-Ls}, \tag{10.1}$$

where $L > 0$ is a pure dead time, and $G_{p0}(s)$ is a strictly proper Hurwitz, minimum-phase transfer function with $G_{p0}(0) \neq 0$.

The control scheme induced from [36] is shown in Figure 10.1(a), where L_m is the estimated dead time, and $G_m(s)$ is a low-order approximation of $G_p(s)$. c, d, n, and y are the command, the disturbance, the measurement noise, and the output, respectively. $\widehat{d}$ is the estimated disturbance. The low-pass filter $Q(s)$ (known as the Q-filter in disturbance observers) is designed to trade-off the robustness and the performance to reject the disturbance and the measurement noise. $C(s)$ is designed according to the delay-free part $G_m(s)$ so that $\frac{C(s)G_m(s)}{1+C(s)G_m(s)}e^{-L_m s}$ meets the desired set-point response. The whole controller consists of two parts: one is the loop $C(s) \sim G_m(s)$, and the other is the disturbance observer of the process. The former serves as a prefilter, and the latter as a feedback loop. The former determines the set-point response, and the latter determines the disturbance response. This control scheme falls into the category of a 2DOF internal model control scheme [76]. The original structure in Figure 10.1(a) is not causal and sometimes not internally stable. An equivalent structure shown in Figure 10.1(b) is causal and internally stable, provided that the relative degree of $Q(s)$ is high enough and the low-pass filter $F(s)$ has a proper relative degree. In the sequel, the controller is designed according to

Figure 10.1(a) but should be implemented according to Figure 10.1(b), where $F(s)$ is used to make the controller proper. It is also necessary to non-dynamically implement the finite-impulse-response block in $\frac{F(s)}{G_m(1-Qe^{-L_m s})}$ to avoid possible pole-zero cancellations [89].

Under the nominal conditions, *i.e.*, when $G_m(s) = G_p(s)$ and $L_m = L$, the transfer functions from the reference command r, the disturbance d, and the measurement noise n to the output y are, respectively,

$$G_{yr}(s) = \frac{C(s)G_m(s)}{1 + C(s)G_m(s)} e^{-L_m s}, \tag{10.2}$$

$$G_{yd}(s) = G_m(s)e^{-L_m s}\left(1 - Q(s)e^{-L_m s}\right), \tag{10.3}$$

$$G_{yn}(s) = Q(s)e^{-L_m s}. \tag{10.4}$$

Obviously, the Smith principle is satisfied, and the dead time is not included in the closed-loop characteristic equation. In addition, $G_{yd}(s) \to 0$ at low frequencies, and $G_{yn}(s) \to 0$ at high frequencies for a low-pass filter $Q(s)$. The system has very good performance to reject disturbances and measurement noises. More importantly, the set-point response is determined by $C(s)$, and the disturbance response is determined by $Q(s)$. In other words, they are *entirely* decoupled from each other.

In general, the controller $C(s)$ may be designed as the proportional controller

$$C(s) = \frac{1}{T_c} \tag{10.5}$$

to obtain the desired set-point response.

10.3 Controller Design to Reject Ramp/Step Disturbances

10.3.1 Design of $Q(s)$

First of all, in order to guarantee the causality of $\frac{Q(s)}{G_m(s)}$ in Figure 10.1(a), the relative degree of $Q(s)$ should be no less than that of $G_m(s)$.

The well-known *internal-model principle* shows that if a disturbance with some modes should be rejected, then the model of the disturbance should be included in the controller. Here, $Q(s)$ can be designed to guarantee the rejection of a known disturbance. Assume that the disturbance polynomial can be represented as $d_d(s)$ with degree n_d and that it has m_d disturbance modes λ_{dk} $(1 \le k \le m_d)$ with multiplicity r_{dk} $(\sum_{k=1}^{m_d} r_{dk} = n_d)$. It is possible [121] to design $Q(s)$ to reject arbitrary disturbances, provided that the disturbance modes λ_{dk} $(1 \le k \le m_d)$ are also the zeros of $1 - Q(s)e^{-L_m s}$. In other words, the disturbance polynomial has to be included in the controller (*implicitly* but not *explicitly*).

In practice, ramp disturbances ($d_d(s) = s^2$) are often used to represent slowly time-varying disturbances. In this case, pole $s = 0$ in the disturbance response (10.3) is of three multiplicities. In other words, $Q(s)$ must be tuned to meet the conditions

$$\begin{cases} 1 - Q(s)e^{-L_m s}|_{s=0} = 0, \\ \frac{d}{ds}\left(1 - Q(s)e^{-L_m s}\right)\Big|_{s=0} = 0, \\ \frac{d^2}{ds^2}\left(1 - Q(s)e^{-L_m s}\right)\Big|_{s=0} = 0, \end{cases} \tag{10.6}$$

or, equivalently,

$$\begin{cases} Q(0) = 1, \\ \dot{Q}(0) = L_m, \\ \ddot{Q}(0) = L_m^2. \end{cases} \tag{10.7}$$

In other words, any low-pass filter with a high enough relative degree that meets the above conditions can be used to reject ramp disturbances. The simplest $Q(s)$ that can meet these conditions is

$$Q(s) = \frac{1 + \mu s + \beta s^2}{(\lambda s + 1)^{d_r+2}}, \tag{10.8}$$

where

$$\mu = (d_r + 2)\lambda + L_m,$$
$$\beta = \frac{L_m^2 + \lambda(d_r + 2)((d_r + 1)\lambda + 2L_m)}{2},$$

and d_r is the relative degree of the delay-free model $G_m(s)$, and λ is a tuning parameter to trade-off the disturbance response and robustness. When implemented as Figure 10.1(b), $F(s)$ may be chosen as $\frac{1}{(\lambda s+1)^{d_r}}$ to guarantee that $\frac{F(s)}{G_m(s)}$ and $\frac{G_m(s)}{F(s)}$ are proper.

For step disturbances, pole $s = 0$ in the disturbance response (10.3) is of two multiplicities. In this case, only the first two conditions in (10.7) should be met. The simplest $Q(s)$ that can meet the conditions is

$$Q(s) = \frac{\mu s + 1}{(\lambda s + 1)^{d_r+1}} \tag{10.9}$$

with

$$\mu = (d_r + 1)\lambda + L_m. \tag{10.10}$$

When the nominal model is $G_m(s) = \frac{K_m}{s}$, then $d_r = 1$. This provides the low-pass filter

$$Q(s) = \frac{(2\lambda + L_m)s + 1}{(\lambda s + 1)^2}, \tag{10.11}$$

and then $F(s)$ can be chosen as $\frac{1}{\lambda s+1}$ without introducing a new parameter. The free parameter λ can be used to compromise disturbance response and robustness.

The loop transfer function of the nominal system is

$$W(s) = \frac{Q(s)e^{-L_m s}}{1 - Q(s)e^{-L_m s}}, \tag{10.12}$$

and the complementary transfer function is

$$T(s) = \frac{W(s)}{1 + W(s)} = Q(s)e^{-L_m s}. \tag{10.13}$$

Hence, for a multiplicative uncertainty $\Delta(s) \in H_\infty$, the system is robustly stable if $\|Q(s)\|_\infty < \frac{1}{\|\Delta(s)\|_\infty}$, *i.e.*, the magnitude frequency response of $Q(s)$ stays beneath that of $\frac{1}{\Delta(s)}$. This can be easily used to guarantee the robust stability using a graphic method.

10.3.2 Examples

10.3.2.1 Example 1: To Reject Step Disturbances

Consider the process studied in [69, 78]

$$G_p(s) = \frac{1}{s(s+1)(0.5s+1)(0.2s+1)(0.1s+1)}, \qquad L = 5 \text{ seconds.}$$

The controller may be designed according to the exact process, which results in a high-order controller. However, the controller proposed in [78] (see Section 8.5) has to be designed according to a reduced model, for example, $\frac{1}{s}e^{-6.5s}$ given in [78].

Here, the controller is designed according to the nominal model $G_m(s) = \frac{1}{s(s+1)(0.5s+1)}$ and $L_m = 5.3$ seconds, where the short-time constants are estimated with a dead time equal to their sum [69]. Hence, $Q(s) = \frac{(4\lambda+5.3)s+1}{(\lambda s+1)^4}$, where $\lambda = 1.3$ is chosen to trade-off the disturbance response with robustness. $C(s)$ is designed as a proportional controller $C(s) = \frac{1}{5}$ to obtain almost the same time constant of the set-point response in [78] with $T_0 = 5$ seconds. The unit-step responses are shown in Figure 10.2 where a step disturbance $d(t) = -0.1$ acts at $t = 70$ seconds. There exists overshoot in the response of Normey-Rico–Camacho scheme discussed in Section 8.5 (noted as N–C in the figures), but there is no overshoot in the response of the disturbance-observer-based scheme (noted as DO in the figures). Moreover, the disturbance response of the disturbance-observer-based scheme is much faster.

When the dead time is $L = 5.5$ sec, the responses under the same controller are shown in Figure 10.3. The performance of the disturbance-observer-based scheme is still much better.

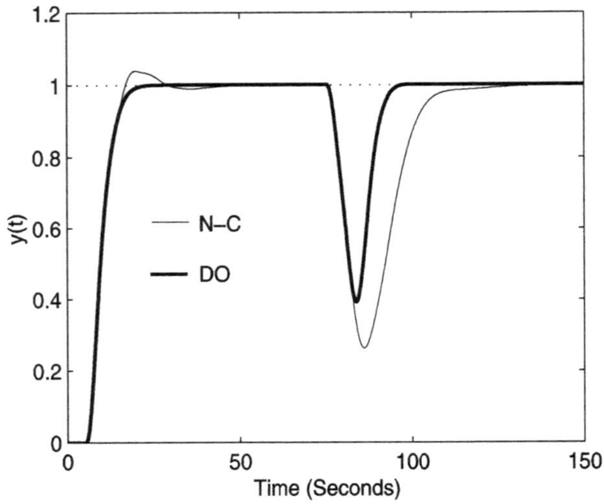

Fig. 10.2 Responses without dead-time uncertainty: step disturbance

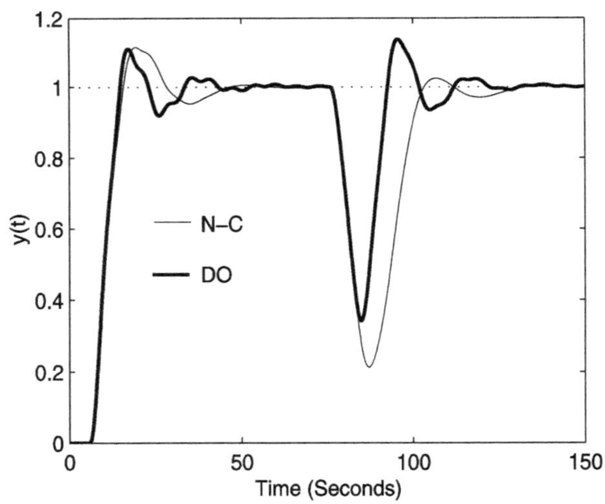

Fig. 10.3 Responses with dead-time uncertainty: step disturbance

10.3.2.2 Example 2: To Reject Ramp Disturbances

Consider the same process as in Example 1. In order to obtain a simpler controller, the proposed controller is designed according to the reduced model $\frac{1}{s}e^{-6.5s}$ as in the case of N–C scheme. The filter is then designed as $Q(s) = \frac{1+\mu s+\beta s^2}{(\lambda s+1)^3}$ with $\lambda = 5$ to meet the robust-stability condition for the additional uncertainty. The system is affected by the ramp disturbance $d(t) = 0.003(t-50)\cdot 1(t-50) - 0.003(t-$

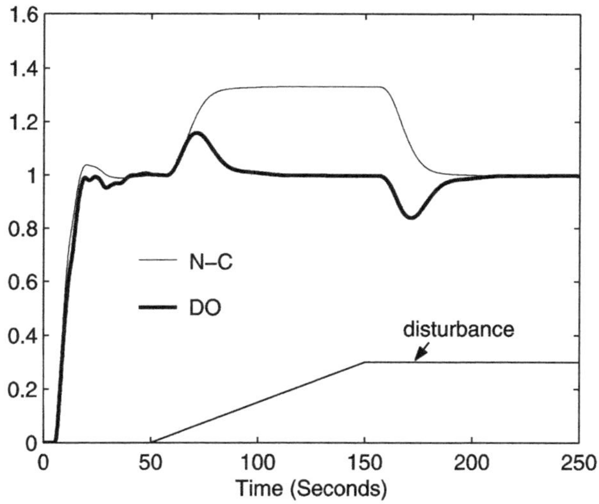

Fig. 10.4 Responses without dead-time uncertainty: ramp disturbance

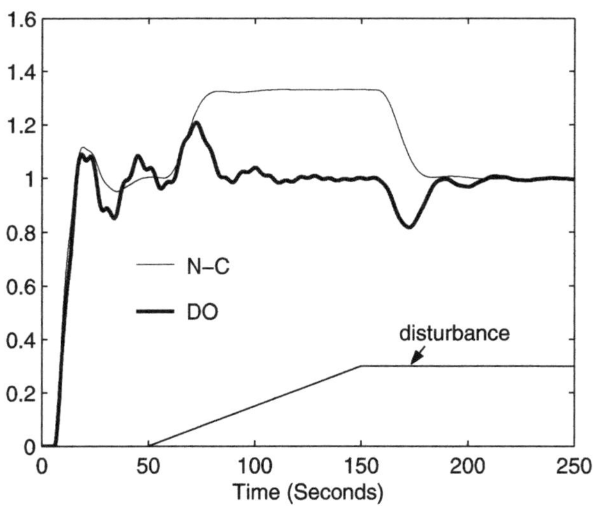

Fig. 10.5 Responses with dead-time uncertainty: ramp disturbance

150) $\cdot$ $1(t-150)$. The responses are shown in Figure 10.4. The proposed system has an excellent capability to reject the *ramp disturbance,* while the scheme in [78] cannot reject the ramp disturbance. As a matter of fact, it is quite difficult to design a controller $C(s)$ so that the N–C scheme has such a capability. When the dead time becomes 5.5 seconds, the responses using the same controller are shown in Figure 10.5. The response of the proposed control scheme is still much better.

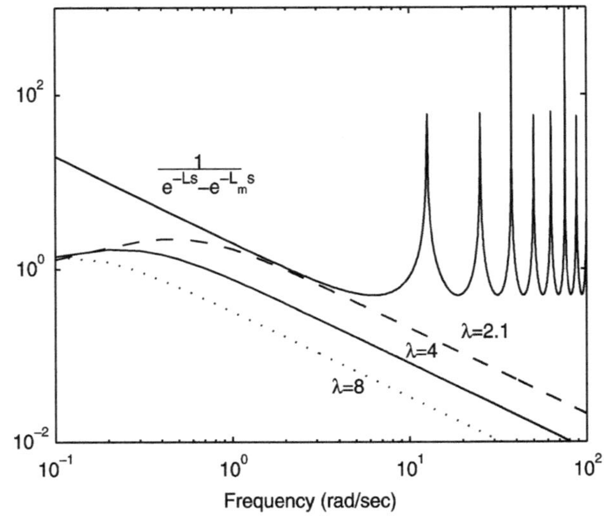

Fig. 10.6 Design of $Q(s)$

10.3.2.3 Example 3

Consider the process with

$$G_m(s) = \frac{1}{s} \quad \text{and} \quad L_m = 5\,\text{sec}, \tag{10.14}$$

where an uncertainty $0 \le L_\Delta \le 0.5$ seconds exists in the dead time $L = L_m + L_\Delta$.

Here, $Q(s) = \frac{(2\lambda+5)s+1}{(\lambda s+1)^2}$ is selected in the form of (10.11). In order to obtain a robustly stable closed-loop system, the bandwidth of $Q(s)$ could not be larger than 3 rad/sec, as shown in Figure 10.6, where three possible candidates of $Q(s)$ with $\lambda = 2.1$, $\lambda = 4$, and $\lambda = 8$ are shown. The robust performance does not degrade considerably for $\lambda = 8$, but the disturbance response is sluggish, as shown in Figures 10.7 and 10.8.

The controller for the set-point response is designed as a P controller $C(s) = 0.5$. The nominal responses of the three cases are shown in Figure 10.7. The system is disturbed by a step disturbance $d_0 = -0.1$ at $t = 50$ seconds. The disturbance response is the best when $Q(s)$ has the broadest bandwidth ($\lambda = 2.1$).

When the dead time varies to the worst case ($L = 5.5$ seconds), the responses under the same controller are shown in Figure 10.8. There exists a limit cycle when $\lambda = 2.1$, and the best robustness is obtained when $\lambda = 8$. Trading-off the disturbance response and the robustness, the best choice of $Q(s)$ can be made as $\lambda = 4$.

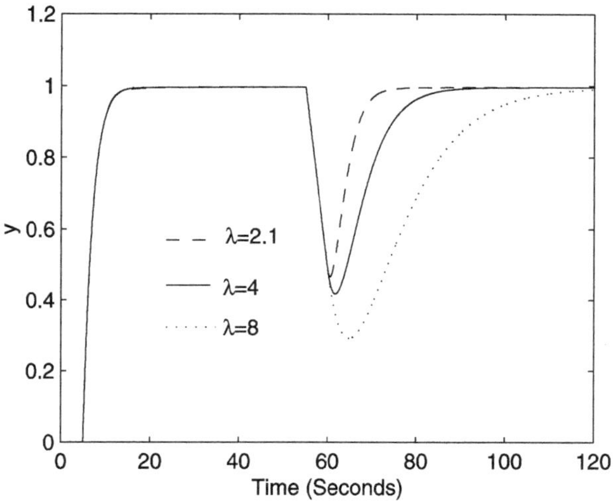

Fig. 10.7 Nominal responses of Example 3

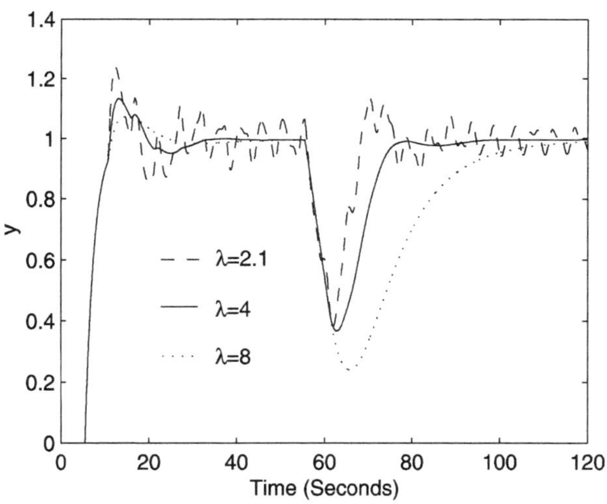

Fig. 10.8 Robust responses of Example 3

10.4 Controller Design to Obtain Deadbeat Disturbance Responses

10.4.1 Design of $Q(s)$

It is possible to design $Q(s)$ to obtain deadbeat responses for step disturbances.

Lemma 10.1 ([77, 146]) *Let $a(s)$ and $b_i(s)$ $(i = 0, 1, \ldots, q)$ be polynomials in s, and $\frac{b_i(s)}{a(s)}$ strictly proper in s. $b(s)$ and $\phi(s)$ are given by*

$$b(s) = b_0(s) + b_1(s)e^{-sT_1} + \cdots + b_q(s)e^{-sT_q} \quad \text{and} \quad \phi(s) = \frac{b(s)}{a(s)},$$

where $T_q > T_{q-1} > \cdots > T_1 > 0$. Let a_j $(j = 1, \ldots, p)$ denote the zeros of $a(s)$ with multiplicity r_j. Assume that $b(s)$ satisfies

$$\left.\frac{d^l}{ds^l} b(s)\right|_{s=a_j} = 0 \quad (l = 0, 1, \ldots, r_j - 1;\ j = 1, \ldots, p). \tag{10.15}$$

Then, $\phi(t) = \mathcal{L}^{-1}[\phi(s)]$ settles to zero in finite time $t = T_q$, i.e.,

$$\phi(t) = \begin{cases} \phi_0(t) & (0 \le t < T_q), \\ 0 & (t \ge T_q). \end{cases}$$

This means that all the zeros of the denominator are also zeros of the numerator. Such a $\phi(s)$ is called a pseudo-differential polynomial [23, 71] because it is actually an entire function of s but not in the exact form of a polynomial. $b(s)$ has infinite zeros because of the multiple delay elements, in addition to the zeros of $a(s)$.

Theorem 10.1 *The disturbance observer-based control system for an integral process with dead time rejects a step disturbance in finite time $t = 2L_m + T_2$ if the low-pass filter is chosen as*

$$Q(s) = \frac{1 - q_1(1 - e^{-T_1 s}) - q_2(1 - e^{-T_2 s})}{\lambda s + 1} \tag{10.16}$$

with

$$\begin{cases} q_1 = -\dfrac{(e^{T_2/\lambda} - 1)(\lambda + L_m) - T_2}{T_2 - T_1 + T_1 e^{T_2/\lambda} - T_2 e^{T_1/\lambda}}, \\ q_2 = \dfrac{(e^{T_1/\lambda} - 1)(\lambda + L_m) - T_1}{T_2 - T_1 + T_1 e^{T_2/\lambda} - T_2 e^{T_1/\lambda}}, \end{cases} \tag{10.17}$$

where $T_2 > T_1 > 0$, and $\lambda > 0$ is a free parameter.

Proof The denominator polynomial of a step disturbance is $D(s) = s$. According to (10.3), the disturbance response is deadbeat if $Q(s)$ is chosen such that the transfer function

$$\phi(s) = G_m(s)\big(1 - Q(s)e^{-L_m s}\big)e^{-L_m s} \cdot \frac{1}{D(s)} \tag{10.18}$$

satisfies Lemma 10.1. $Q(s)$ in (10.16) is such a candidate. Substitute (10.16) into (10.18); then

$$\phi(s) = \frac{\lambda s + 1 - e^{-L_m s} + q_1(1 - e^{-T_1 s})e^{-L_m s} + q_2(1 - e^{-T_2 s})e^{-L_m s}}{s^2(\lambda s + 1)} k e^{-L_m s}. \tag{10.19}$$

The poles of $\phi(s)$ are $0,\ 0,$ and $-\frac{1}{\lambda}$; $\phi(s)$ also has a zero at $s=0$. Hence, in order to make $\phi(s)$ be a pseudo-differential polynomial, the following equations should be met according to Lemma 10.1:

$$\begin{cases} -q_1T_1 - q_2T_2 - \lambda = L_m, \\ 1 - q_1\left(1-e^{T_1/\lambda}\right) - q_2\left(1-e^{T_2/\lambda}\right) = 0. \end{cases} \tag{10.20}$$

For $T_2 > T_1 > 0$, the solutions of q_1 and q_2 can be obtained as (10.17). This completes the proof. □

For any tuning parameters $T_2 > T_1 > 0$ and $\lambda > 0$, the static gain of $Q(s)$ given in (10.16) is always equal to 1. This guarantees the stability of the closed-loop system.

The filter $Q(s)$ may be interpreted as a low-pass filter, $\frac{1}{\lambda s+1}$, cascaded with an *input shaper* [12, 90, 164], $1-q_1-q_2+q_1e^{-T_1s}+q_2e^{-T_2s}$. For an arbitrary disturbance polynomial $D(s)$, it is possible to obtain a deadbeat response, provided that the number of the delay elements is enough so that $\phi(s)$ in (10.18) is a pseudo-differential polynomial.

10.4.2 Implementation of the Controller

In order to make the measurement unit $\frac{Q(s)}{F(s)}$, the input filter $\frac{CG_m}{(1+CG_m)F(s)}$, and the controller $\frac{F(s)}{G_m(1-Qe^{-L_ms})}$ in Figure 10.1(b) *causal*, the relative degree of $F(s)$ has to be 1 for the above design. The simplest $F(s)$ is to choose

$$F(s) = \frac{1}{\lambda s+1},$$

without introducing a new parameter. Then,

$$\frac{CG_m}{(1+CG_m)F(s)} = \frac{\lambda s+1}{Ts+1},$$

$$\frac{Q(s)}{F(s)} = 1 - q_1\left(1-e^{-T_1s}\right) - q_2\left(1-e^{-T_2s}\right),$$

and

$$\begin{aligned} &\frac{F(s)}{G_m(1-Qe^{-L_ms})} \\ &= \frac{s}{K(\lambda s+1-e^{-L_ms}+q_1(1-e^{-T_1s})e^{-L_ms}+q_2(1-e^{-T_2s})e^{-L_ms})} \\ &= \frac{1}{K}\cdot\frac{1}{\lambda+\frac{1-e^{-L_ms}}{s}+q_1\frac{1-e^{-T_1s}}{s}e^{-L_ms}+q_2\frac{1-e^{-T_2s}}{s}e^{-L_ms}} \\ &= \frac{1}{K}\cdot\frac{\frac{1}{\lambda}}{1+\frac{1}{\lambda}Z(s)}, \end{aligned}$$

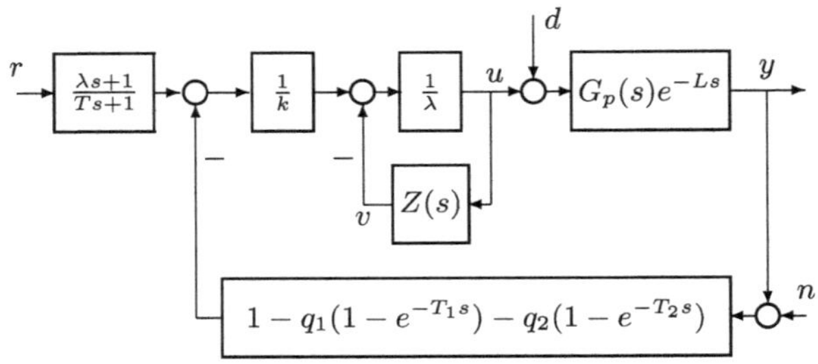

Fig. 10.9 The control structure for implementation

where

$$Z(s) \doteq \frac{1-e^{-L_m s}}{s} + q_1 \frac{1-e^{-T_1 s}}{s} e^{-L_m s} + q_2 \frac{1-e^{-T_2 s}}{s} e^{-L_m s}$$

is an FIR block or a pseudo-differential polynomial. Hence, the control system can be implemented as the structure shown in Figure 10.9, where the FIR block $Z(s)$ has to be implemented in a non-dynamical way [89], *i.e.*, to implement the signal v in time-domain as

$$v(t) = \int_{t-L_m}^{t} u(t)\,dt + q_1 \int_{t-L_m-T_1}^{t-L_m} u(t)\,dt + q_2 \int_{t-L_m-T_2}^{t-L_m} u(t)\,dt.$$

By doing so, $Z(s)$ does not possess any pole, even at $s=0$, and there does not exist any unstable pole-zero cancellation between the controller and the plant. Hence, the control scheme implemented in Figure 10.9 is internally stable.

In this control structure, the disturbance response from d to y, considering uncertainties in the plant, can be represented as

$$\begin{aligned} G_{yd}(s) &= \frac{\frac{K}{s} G_{p0}(s) e^{-Ls}}{1+(1-q_1(1-e^{-T_1 s})-q_2(1-e^{-T_2 s}))\frac{1}{\lambda+Z(s)}\frac{1}{s}G_{p0}(s)e^{-Ls}} \\ &= \frac{K(\lambda+Z(s))G_{p0}(s)e^{-Ls}}{\lambda s + sZ(s) + (1-q_1(1-e^{-T_1 s})-q_2(1-e^{-T_2 s}))G_{p0}(s)e^{-Ls}}. \end{aligned}$$

Since $G_{p0}(0) \neq 0$ and $Z(s)$ does not have a pole, $G_{yd}(s)$ does *not* have an unstable pole at $s=0$. In the nominal case, where $G_{p0}(s)=1$ and $L_m = L$, the disturbance response is

$$G_{yd}(s) = \frac{K\,(\lambda + Z(s))\,e^{-Ls}}{\lambda s + 1}. \tag{10.21}$$

There is only one pole at $s=-\frac{1}{\lambda}$ (since $Z(s)$ does not possess a pole) and the nominal system is stable. Moreover, this (stable) pole is cancelled by the zero at $s=-\frac{1}{\lambda}$ since $\lambda + Z(-\frac{1}{\lambda}) = 0$, according to (10.20). This confirms that the obtained disturbance response is indeed deadbeat in the nominal case.

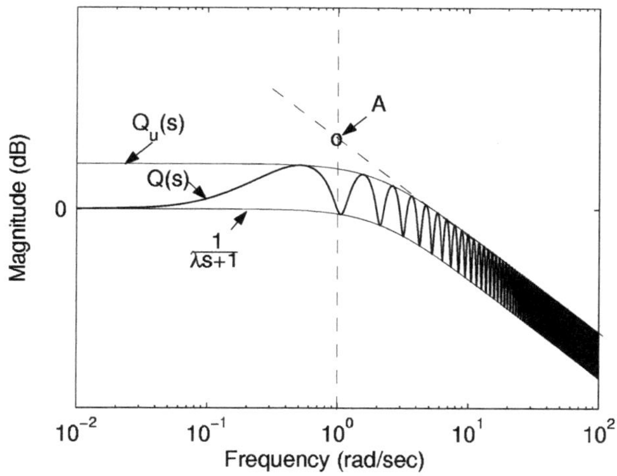

Fig. 10.10 Magnitude frequency response of $Q(s)$

10.4.3 Parameter Tuning and Robustness

The complementary transfer function of the nominal system is

$$T(s) = Q(s)e^{-L_m s}.$$

Hence, for a multiplicative uncertainty $\Delta(s) \in H_\infty$, the system is robustly stable if $\|Q(s)\|_\infty < \frac{1}{\|\Delta(s)\|_\infty}$, *i.e.*, the magnitude frequency response $|Q(j\omega)|$, of which a typical example is shown in Figure 10.10, stays under $\frac{1}{|\Delta(j\omega)|}$.

It can be shown that for $T_2 > T_1 > 0$ and $\lambda > 0$, the (delay) gains satisfy

$$1 - q_1 - q_2 > 0, \qquad q_1 < 0, \quad \text{and} \quad q_2 > 0. \tag{10.22}$$

The upper envelope of $Q(s)$ given in (10.16) is

$$Q_u(s) = \frac{|1 - q_1 - q_2| + |q_1| + |q_2|}{\lambda s + 1} = \frac{1 - 2q_1}{\lambda s + 1}.$$

Hence, the following theorem holds:

Theorem 10.2 *The system described in Theorem* 10.1 *is robustly stable if*

$$1 - 2q_1 < \frac{1}{\left\| \frac{\Delta(s)}{\lambda s + 1} \right\|_\infty}$$

for a multiplicative uncertainty $\Delta(s) \in H_\infty$.

The high-frequency asymptote of the upper envelope $Q_u(s)$ is

$$\log\left|Q_u^a(j\omega)\right| = \log \frac{1 - 2q_1}{\lambda} - \log \omega.$$

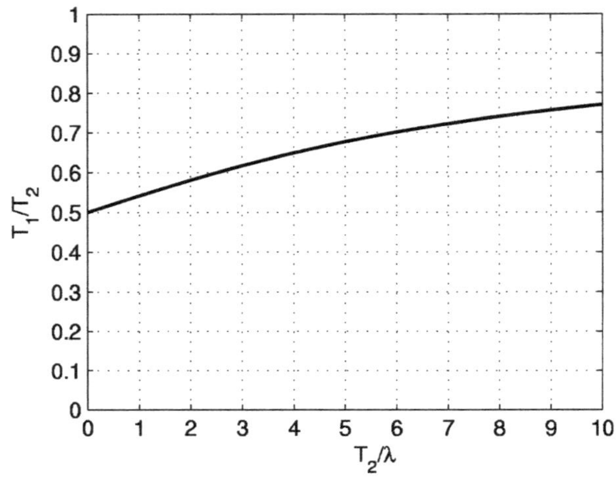

Fig. 10.11 The optimal ratio T_1/T_2 with respect to T_2/λ

It intersects with the straight line $\omega = 1$ rad/sec at $\log \frac{1-2q_1}{\lambda}$, noted as A in Figure 10.10. $1-2q_1$ can be interpreted as the *cost of the robustness to obtain a deadbeat disturbance response*, and $J = \frac{1-2q_1}{\lambda}$ as *a robustness indicator*. For given T_2 and λ, T_1 can be chosen to make point A as low as possible, *i.e.*,

$$\min_{T_1} J = \min_{T_1} \frac{1-2q_1}{\lambda},$$

to obtain the largest possible robust stability region. With consideration of (10.17) and (10.22), the robustness indicator can be simplified as

$$J = \frac{1}{\lambda}\left(1 + \frac{2(\lambda + L_m)(e^{T_2/\lambda} - 1) - 2T_2}{T_2 - T_1 + T_1 e^{T_2/\lambda} - T_2 e^{T_1/\lambda}}\right). \tag{10.23}$$

Since $(\lambda + L_m)(e^{T_2/\lambda} - 1) - T_2 > 0$ and $T_2 - T_1 + T_1 e^{T_2/\lambda} - T_2 e^{T_1/\lambda} > 0$ for $T_2 > T_1 > 0$ and $\lambda > 0$, J is always larger than $\frac{1}{\lambda}$. Differentiate J with respect to T_1 and let it be 0; then

$$-1 + e^{T_2/\lambda} - \frac{T_2}{\lambda} e^{T_1/\lambda} = 0,$$

which offers

$$T_1 = \lambda \ln \frac{e^{T_2/\lambda} - 1}{T_2/\lambda}, \tag{10.24}$$

or

$$\frac{T_1}{T_2} = \frac{\lambda}{T_2} \ln \frac{e^{T_2/\lambda} - 1}{T_2/\lambda}.$$

A graphical representation of this function is shown in Figure 10.11.

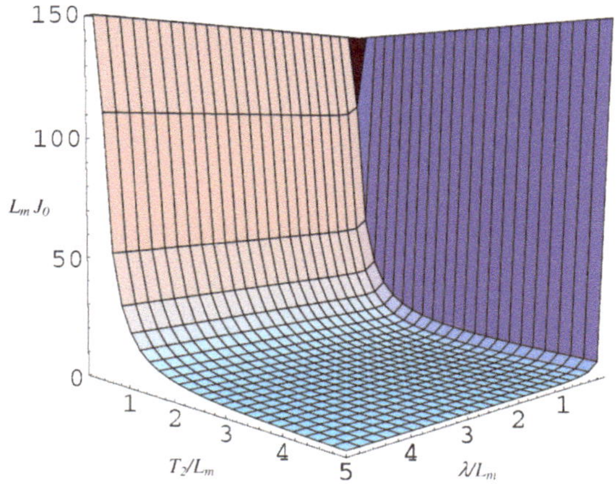

Fig. 10.12 Scaled robustness indicator $L_m J_o$

If $T_2/\lambda \to 0$, then $T_1 \to 0.5T_2$; If $T_2/\lambda \to \infty$, then $T_1 \to T_2$. Thus, T_1 is always less than T_2. The corresponding minimal J is then obtained as

$$J_o = \frac{1}{\lambda}\left(1 + \frac{2\left(1+\frac{L_m}{\lambda}\right)(e^{T_2/\lambda}-1) - 2\frac{T_2}{\lambda}}{\frac{T_2}{\lambda} + (e^{T_2/\lambda}-1)\left(\ln\frac{e^{T_2/\lambda}-1}{T_2/\lambda} - 1\right)}\right). \tag{10.25}$$

A scaled J_o by L_m is shown in Figure 10.12 with respect to $\frac{T_2}{L_m}$ and $\frac{\lambda}{L_m}$.

As can be seen, the larger T_2 and λ, the better the robustness. The robustness becomes worse for $T_2 < L_m$ or very small λ/L_m. Hence, T_2/L_m or λ/L_m cannot be chosen too small, although a small λ is helpful for the dynamics of the disturbance response as given in (10.21).

If $\lambda \to +\infty$, then $L_m J_o \to \frac{16+8\frac{T_2}{L_m}}{(\frac{T_2}{L_m})^2}$. In fact, for $\lambda > L_m$, the robustness indicator J_o varies slightly. This means that λ can no longer be used as a fine tuning parameter to meet the robustness as in a common disturbance observer-based control scheme using a rational low-pass filter [58, 163]. A reasonable value of λ is $(0.5 \sim 1)L_m$.

If $T_2 \to +\infty$, then $J_o \to \frac{1}{\lambda}$; the system degenerates into a common system without the property of a deadbeat disturbance response and so the robustness. This means that T_2 has to be used to compromise between the deadbeat time and the robustness. If a deadbeat disturbance response is desired, the robustness has to be sacrificed to some extent. If good robustness is desired, then the deadbeat time cannot be chosen too short. This also indicates that delay T_2 takes the place of the bandwidth of the low-pass filter to compromise the disturbance response with robustness. This compromise can never be overcome.

For given T_2 and λ, J_o is proportional to the dead time L_m as shown in (10.23). This means that a system having a longer dead time has to pay more for the robustness in order to obtain a deadbeat disturbance response.

In summary, the parameters are determined as follows:

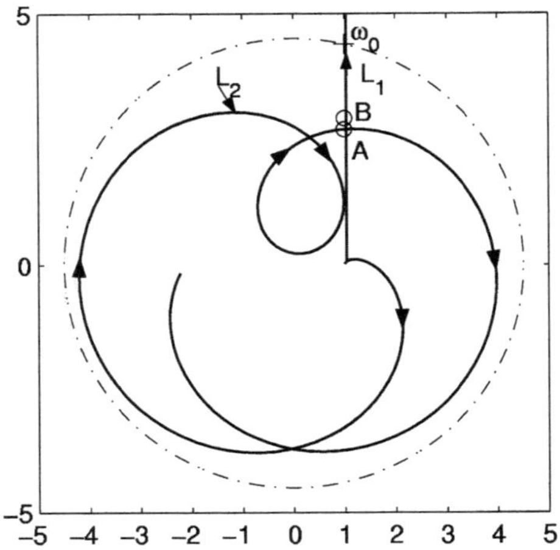

Fig. 10.13 Dual-locus diagram of the controller

Step 1. Initially choose $T_2 = (1.5 \sim 2)L_m$; then the recovery time is $(3.5 \sim 4)L_m$, which is much faster than that of the sub-ideal disturbance response $(5.5 \sim 9)L_m$ (if the robustness is met);

Step 2. Choose $\lambda = (0.5 \sim 1)L_m$, calculate T_1 from (10.24), and then q_1 and q_2 from (10.17).

Step 3. Evaluate the robustness. If the robust stability condition is not met, then increase λ and repeat Step 2 to check if the robust stability condition can be slightly over-met; if the robust stability condition is much over-met, then decrease T_2 and repeat Step 2 till the robust stability condition is slightly over-met. If increasing λ cannot meet the robust stability condition, then increase T_2 and repeat the above procedures again till the robust stability condition is slightly over-met.

10.4.4 Stability of the Controller

The following analysis shows that the controller itself has no right-half-plane pole and hence is stable. The characteristic equation of the controller itself, $1 - Q(s)e^{-L_m s} = 0$, is equivalent to

$$\lambda s + 1 = (1 - q_1 - q_2)e^{-L_m s} + q_1 e^{-(T_1+L_m)s} + q_2 e^{-(T_2+L_m)s}.$$

Denote the left side as $L_1(s)$ and the right side as $L_2(s)$, of which the loci are shown in Figure 10.13 for $\omega = 0 \to +\infty$. Locus $L_1(j\omega)$ is a straight line from $(1,\ j0)$ to $(1,\ j\infty)$, and locus $L_2(j\omega)$ lies inside the circle at $(0,0)$ with a radius of $1 - 2q_1 = 4.5$. $L_1(j\omega)$ intersects with the circle at $\omega_0 = \frac{\sqrt{(1-2q_1)^2-1}}{\lambda} = 1.75$ rad/sec. For $\omega >$

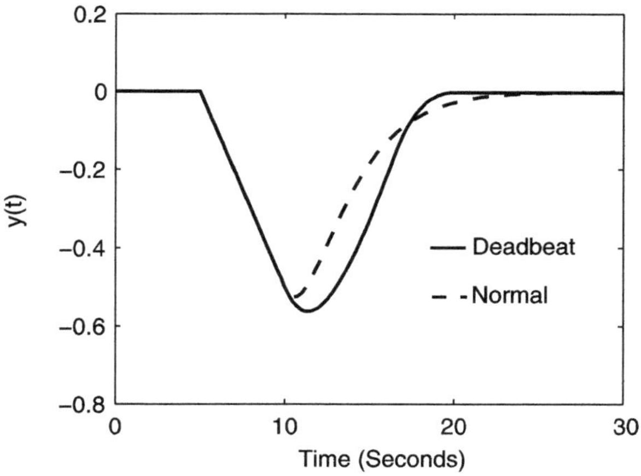

(a) The nominal case

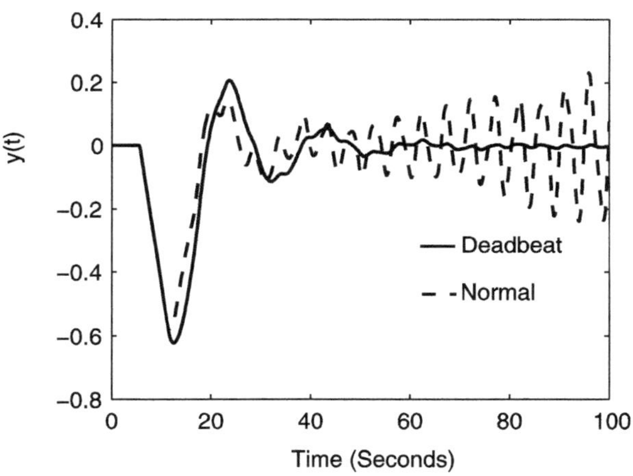

(b) The worst case of uncertainty

Fig. 10.14 Comparison of disturbance responses

ω_0, the two loci do not intersect. For $0 < \omega < \omega_0$, there is only one intersection (noted as A in Figure 10.13), of which the corresponding frequency of $L_2(j\omega)$ is about $\omega_A = 1.165$ rad/sec. For the same frequency, $L_1(j\omega_A)$ arrives at point B. Hence, L_1 arrives at the intersection A earlier than L_2. According to the dual-locus diagram [114, 157], no right-half-plane pole exists in the controller, and, hence, the controller itself is stable.

10.4.5 An Example

Consider again the process with

$$G_m(s) = \frac{1}{s}, \qquad L_m = 5\,\text{seconds},$$

assuming that the worst multiplicative uncertainty is $\Delta(s) = \frac{1}{0.1s+1}e^{-0.5s} - 1$.

Choose $T_2 = 2L_m = 10$ seconds and $\lambda = 0.5L_m = 2.5$ seconds; then $T_1 = 6.5$ seconds, $q_1 = -1.75$, $q_2 = 0.39$. The disturbance response of $d(t) = -0.1$ in the nominal case is shown in Figure 10.14(a), and the one in the worst case of uncertainty is shown in Figure 10.14(b). The nominal disturbance response is deadbeat at $t = 2L_m + T_2 = 20$ seconds, and the response in the worst case uncertainty becomes worse but still very good. A comparison to the case described in the previous section with $Q(s) = \frac{9s+1}{(2s+1)^2}$ to obtain a recovery time close to $t = 20$ seconds is made. The corresponding disturbance responses in the nominal and worst cases are noted as Normal in Figures 10.14 (a) and (b), respectively. As can be seen, the system becomes unstable in the worst case of uncertainty. In other words, in order to obtain the same recovery time, the proposed deadbeat control scheme has much better robustness than the conventional scheme. In order to obtain similar robustness, the recovery time obtained by the conventional scheme is much longer. The advantage of the conventional schemes is a relatively smaller dynamic error.

10.5 Conclusions

A disturbance observer-based control scheme, which is a 2DOF internal model control structure, is revealed to be very effective to control IPDT processes. It has a simple, clear, easy-to-design, and easy-to-implement structure; it decouples the set-point response from the disturbance response; it can be designed to reject arbitrary disturbances. The controller can be designed to reject arbitrary disturbances, and it is possible to obtain deadbeat disturbance responses for step disturbances.

Chapter 11
Quantitative Analysis

As shown in the previous chapters, several different control schemes for integral processes with dead time resulted in the same disturbance response. Moreover, it has already been shown that such a response is sub-ideal [73, 160]. In this chapter, the achievable specifications of this disturbance response and the robust stability regions of the system are quantitatively analysed. The control parameter is quantitatively determined with compromise between the disturbance response and the robustness. Four specifications—(normalised) maximal dynamic error, maximal decay rate, (normalised) control action bound, and approximate recovery time—are given to characterise the step-disturbance response. It shows that any attempt to obtain a (normalised) dynamic error less than L_m is impossible, and a sufficient condition on the (relative) gain-uncertainty bound is $\frac{\sqrt{3}}{2}$.

11.1 Introduction

It is very interesting that, although the structures proposed in [78, 153] (see Sections 8.3.3 and 8.5.3, respectively) are different from the control scheme presented in [163] (see Chapter 10), the resulted disturbance responses are the same in essence when the controllers are properly tuned. Further research has shown that such a response is sub-ideal [72]. Hence, it is quite necessary to quantitatively analyse the system performance with respect to the control parameter. The relationship between the control parameter and the achievable specifications and the robust stability regions is shown in this chapter, based on the disturbance observer-based control scheme presented in Chapter 10. The control parameter can be decided by a given allowable maximal dynamic error, in terms of the *Lambert W* function [17]. For a given control parameter, a control action bound is required. The maximal decay rate, based on which the recovery time is approximately obtained, is given to measure the responding speed to the disturbance. For a given gain uncertainty, dead-time uncertainty, or both of them, the corresponding stability region can be derived. Hence, the control parameter can be analytically determined with compromise be-

A. Visioli, Q.-C. Zhong, *Control of Integral Processes with Dead Time*,
Advances in Industrial Control,
DOI 10.1007/978-0-85729-070-0_11, © Springer-Verlag London Limited 2011

tween robustness and the disturbance response. A sufficient condition on the relative gain-uncertainty bound is $\frac{\sqrt{3}}{2}$.

11.2 The *Lambert W* Function

The *Lambert W* function is named after a great mathematician, *Johann Heinrich Lambert* in the 18th century, and also in memory of the pioneering work on this function by E.M. Wright. In some literature, it is also called the Ω-function.

It is defined to be the multi-valued inverse of a very simple function $x = we^w$. However, it is a very interesting function not only in the eyes of a mathematician. It has been widely used in, for example, the enumeration of trees, the calculation of water-wave heights, the solution of linear delay equations with constant coefficients, and so on. The history of this function and applications can be found in [17] and the references therein.

The function W_k ($k \in \mathbb{Z}$) may be defined for $x \in \mathbb{C}\backslash(-\infty, 0]$ as the (unique) solution of

$$W_k(x) = \ln x + j2k\pi - \ln W_k(x), \tag{11.1}$$

where ln is the principal branch of the logarithm. Note that (11.1) implies that if $w = W_k(x)$, then $we^w = x$, which is the basic equation satisfied by all the branches W_k. For $x > 0$, $W_0(x)$ is the only positive solution of $we^w = x$. Equation (11.1) can be thought of as a fixed-point equation for the function

$$T_{k,x}(w) = \ln x + j2k\pi - \ln w.$$

For $k \neq 0$ and large $|x|$, the iterations defined by $w_{n+1} = T_{k,x}(w_n)$ converge fast to $W_k(x)$. The approximation formula (4.11) given in [17] can be written as $W_k(x) \approx T_{k,x}(T_{k,x}(1))$. Since for large real x, $W_0(x)$ is a better initial approximation of $W_k(x)$ than 1, the approximation

$$W_k(x) \approx T_{k,x}\big(T_{k,x}\big(W_0(x)\big)\big) = \ln x + j2k\pi - \ln\big(\ln x + j2k\pi - \ln W_0(x)\big) \tag{11.2}$$

can be used.

The principle branch $W(x) = W_0(x)$ of the *Lambert W* function is shown in Figure 11.1(a) for $-e^{-1} \leq x < 0$ and in Figure 11.1(b) for $0 \leq x \leq e$.

11.3 Achievable Specifications of the Sub-ideal Disturbance Response

The integral processes with dead time can be described as

$$G(s) = G_p(s)e^{-Ls} = \frac{K}{s}e^{-Ls}, \tag{11.3}$$

where the pure dead time L and the static gain K are both positive.

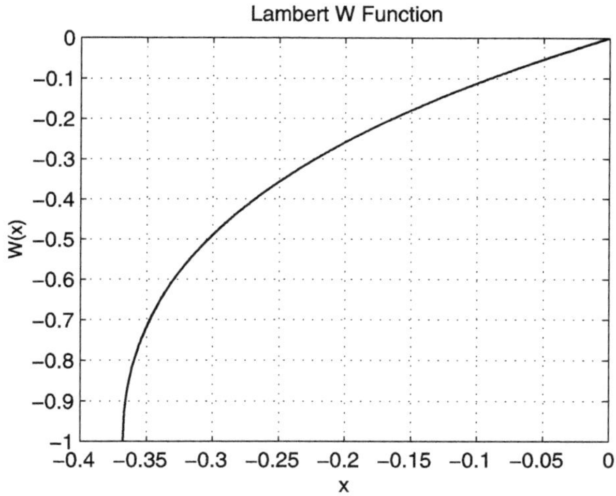

(a) $-e^{-1} \leq x < 0$

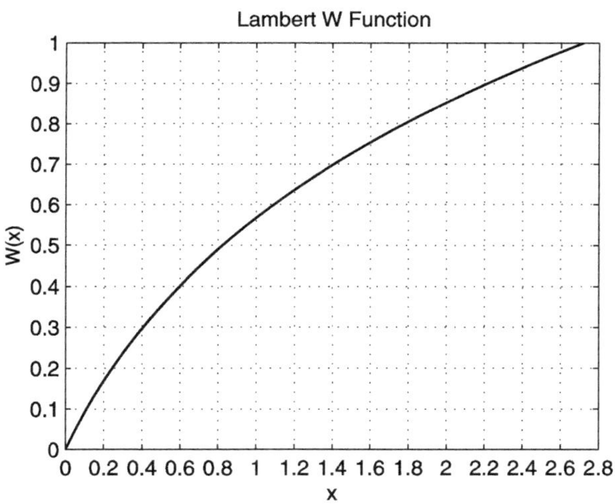

(b) $0 \leq x \leq e$

Fig. 11.1 The principle branch of the *Lambert W* function

The disturbance observer-based control scheme is revisited in Figure 11.2(a) for the readers' convenience. L_m is the estimated dead time, and $G_m(s)$ is the nominal delay-free part. $C(s)$ is designed as the proportional controller $C(s) = \frac{1}{KT}$, and $Q(s)$ is the low-pass filter

$$Q(s) = \frac{(2\lambda + L_m)s + 1}{(\lambda s + 1)^2} \tag{11.4}$$

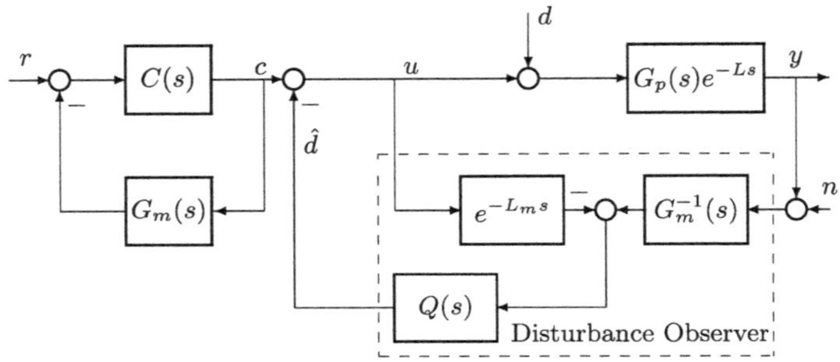

(a) Original structure for design

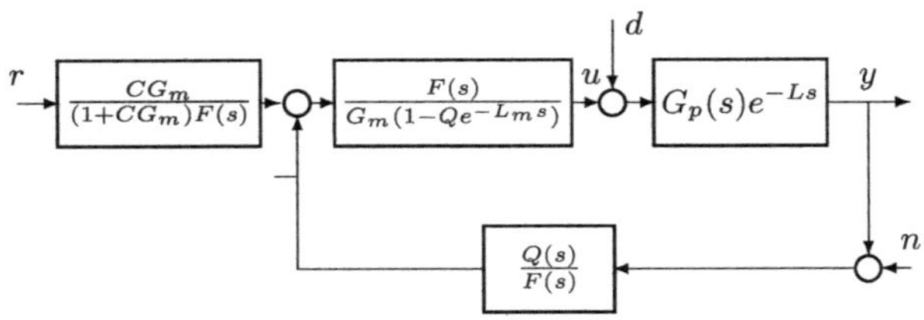

(b) Equivalent structure for implementation

Fig. 11.2 Disturbance observer-based control scheme

designed to reject step disturbances without a static error, where λ is a design parameter to trade-off the disturbance rejection, robustness, and, possibly, the control action bound. A casual and internally stable, equivalent structure is shown in Figure 11.2(b).

Under the nominal condition, $G_m(s) = G_p(s) = \frac{K}{s}$ and $L_m = L$, the set-point response and the disturbance response are respectively

$$G_{yr}(s) = \frac{1}{Ts+1}e^{-L_m s},$$

$$G_{yd}(s) = \frac{k}{s}\left(1 - Q(s)e^{-L_m s}\right)e^{-L_m s}. \qquad (11.5)$$

It is quite easy to evaluate the achievable specifications of the set-point response. However, it is not that easy to evaluate the achievable specifications of the disturbance response. The disturbance response obtained using (11.4) is sub-ideal [72]. A typical step disturbance response is shown in Figure 11.3. Four specifications—maximal dynamic error, maximal decay rate, control action bound and approximate recovery time—will be given to characterise the disturbance response.

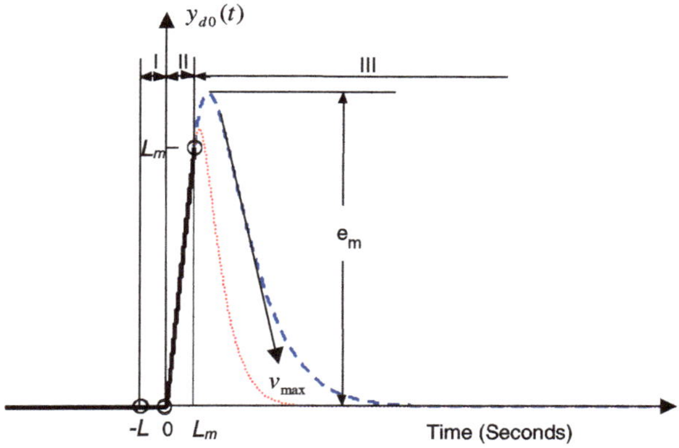

Fig. 11.3 The normalised response of a step disturbance

11.3.1 Maximal Dynamic Error

Theorem 11.1 *For process* (11.3) *controlled as in Figure* 11.2, *the* (*normalised*) *maximal dynamic error*[1] e_m *under a step disturbance is not less than* L_m, i.e., $e_m \geq L_m$.

Proof The normalised response, $y_{d0}(t) = \frac{y_d(t)}{d_0 K}$, of a step disturbance d_0, disturbing at $t_0 = -L$ without loss of generality is

$$y_{d0}(t) = \mathcal{L}^{-1}\left[\frac{1}{s^2}\left(1 - Q(s)e^{-L_m s}\right)\right]$$
$$= t - \mathcal{L}^{-1}\left[\frac{1}{s^2}Q(s)\right] \cdot 1(t - L_m). \qquad (11.6)$$

$y_{d0}(t)$ can be divided into the following three stages as shown in Figure 11.3:

Stage I: the output is not affected due to the effect of the dead time in the process;
Stage II: after dead time L the step disturbance acts, the output responds proportionally to time t, during which the control action is delayed by the dead time;
Stage III: after delay $L + L_m$, the control action starts to affect the output.

The first two stages are not controllable. At the end of Stage II, the dynamic error reaches L_m. After that, the dynamic error may increase or decrease. Whatever it varies, the normalised maximal dynamic error is not less than L_m. This completes the proof. □

Clearly, it is impossible to obtain a dynamic error that is less than L_m.

Substituting Equation (11.4) into (11.6), the normalised response of a step disturbance can be obtained as

[1] e without a subscript stands, as usual, for the exponential constant.

$$y_{d0}(t) = t - \mathcal{L}^{-1}\left[\frac{(2\lambda + L_m)s + 1}{s^2(\lambda s + 1)^2}\right] \cdot 1(t - L_m)$$
$$= \begin{cases} t & (0 \le t \le L_m), \\ (\frac{\lambda + L_m}{\lambda} t - \frac{L_m^2}{\lambda}) e^{-\frac{t - L_m}{\lambda}} & (t \ge L_m). \end{cases} \tag{11.7}$$

Theorem 11.2 (Allowable Control Parameter λ) *If process* (11.3) *is controlled as in Figure* 11.2 *and the low-pass filter is chosen in the form of Equation* (11.4), *then for a given normalised maximal dynamic error* $e_m > L_m$, *the control parameter* λ *should satisfy the condition*

$$0 < \lambda \le \frac{L_m}{-W\left(-\frac{L_m}{e_m e}\right)} - L_m, \tag{11.8}$$

where $W(\cdot)$ *is the* Lambert *W function.*

Proof As stated in Theorem 11.1, the output can be affected only when $t > L_m$. Hence, there is no need to consider the first two stages of the response. The response in Stage III can be represented as

$$y_{d03}(t) = \left(\frac{\lambda + L_m}{\lambda} t - \frac{L_m^2}{\lambda}\right) e^{-\frac{t - L_m}{\lambda}}. \tag{11.9}$$

$y_{d03}(t)$ reaches the peak value, *i.e.*, the maximal dynamic error e_m:

$$e_m = y_{d03}(t_m) = (\beta + 1) L_m e^{-\frac{\beta}{\beta + 1}} \tag{11.10}$$

(with $\beta \doteq \lambda / L_m$) at the moment

$$t_m = \frac{\lambda^2 + \lambda L_m + L_m^2}{\lambda + L_m} = L_m + \frac{\beta^2}{\beta + 1} L_m.$$

Solve β from (11.10); then

$$\beta = \frac{1}{-W\left(-\frac{L_m}{e_m e}\right)} - 1 > 0.$$

Differentiating Equation (11.10) with respect to β or λ shows that the maximal dynamic error e_m is an increasing function with respect to λ. The larger the λ, the larger the e_m. This completes the proof. □

11.3.2 Maximal Decay Rate

Differentiating Equation (11.9) with respect to t twice and letting it be 0, the maximal decay rate $v_{\max}$ can be obtained as

$$v_{\max} = -\frac{\lambda + L_m}{\lambda e} e^{-\frac{\lambda}{\lambda + L_m}} \tag{11.11}$$

at the moment $t_{vm} = \lambda + L_m + \frac{\lambda^2}{\lambda+L_m}$. $v_{\max}$ is an increasing function with respect to λ and can be served as a specification of the recovery time. The larger the $|v_{\max}|$, the shorter the recovery time, and the faster the disturbance response. There exists a minimum value $|v_{\max}|_{\min} = e^{-2}$ as $\lambda \to \infty$. This is the slowest one.

11.3.3 Control Action Bound

The bound of the control action can be characterised by the theorem below.

Theorem 11.3 *If process* (11.3) *is controlled by the above proposed method and the low-pass filter is chosen in the form of Equation* (11.4), *then a* (*normalised*) *control action bound* $1 + \frac{\lambda+L_m}{\lambda e} e^{-\frac{\lambda}{\lambda+L_m}}$ *is required. In other words, for a given normalised control action bound* u_m, *the control parameter should satisfy the following condition*:

$$\lambda \geq \frac{W\left(\frac{1/e}{u_m-1}\right)}{1 - W\left(\frac{1/e}{u_m-1}\right)} L_m. \tag{11.12}$$

Proof The transfer function from disturbance d to control action u under the nominal case is

$$G_{ud}(s) = Q(s)e^{-Ls}.$$

Assume that a step disturbance d_0, disturbing at $t_0 = -L$ without loss of generality, then the normalised control action, $u_0(t) = \frac{u(t)}{d_0}$, for $t \geq 0$ is

$$\begin{aligned} u_0(t) &= \mathcal{L}^{-1}\left[\frac{1}{s}Q(s)\right] \\ &= 1 + \left(\frac{L_m t}{\lambda^2} + \frac{t}{\lambda} - 1\right)e^{-\frac{t}{\lambda}}. \end{aligned}$$

$u_0(t)$ reaches the peak value, *i.e.*, the normalised control action bound, u_m, is

$$u_m = 1 + \frac{\lambda + L_m}{\lambda e} e^{-\frac{\lambda}{\lambda+L_m}} \tag{11.13}$$

at the moment $t_{um} = \lambda + \frac{\lambda^2}{\lambda+L_m}$. For a given normalised control action bound u_m, the control parameter λ can be solved as

$$\lambda = \frac{W\left(\frac{1/e}{u_m-1}\right)}{1 - W\left(\frac{1/e}{u_m-1}\right)} L_m.$$

It can be shown that u_m is a decreasing function with respect to λ. Hence, for a given control bound $u_m > 0$, the control parameter should satisfy condition (11.12). This completes the proof. □

For the minimal maximal dynamic error $e_m \to L_m$, there should be $\lambda \to 0$ and $u_m \to \infty$. Hence, it is impossible to obtain a dynamic error equal to L_m even under the nominal case because of the infinite control action. This means that only a sub-ideal disturbance response can be obtained. There exists a low bound for u_m: $u_m \to 1 + e^{-2} \approx 1.14$ as $\lambda \to \infty$.

According to Theorems 11.2 and 11.3, the following results are obtained:

Corollary 11.1 *For a given normalised maximal dynamic error $e_m > L_m$, a control action bound $u_m = 1 + \frac{e_m}{\lambda e}$ is required, where λ is limited by Theorem 11.2.*

Corollary 11.2 *For a given normalised maximal dynamic error $e_m > L_m$ and a normalised control action bound u_m, the control parameter should meet the following condition:*

$$\frac{W\left(\frac{1/e}{u_m - 1}\right)}{1 - W\left(\frac{1/e}{u_m - 1}\right)} L_m \le \lambda \le \frac{L_m}{-W\left(-\frac{L_m}{e_m e}\right)} - L_m. \tag{11.14}$$

11.3.4 Approximate Recovery Time

Combining Equations (11.10) and (11.11), the following relation between e_m and $v_{\max}$ holds:

$$e_m = -\lambda e v_{\max}.$$

This means that if the response decays at the rate $v_{\max}$, then it decays to 0 after λe from the peak time t_m. Assume that the average decay rate is half of the maximal decay rate, which is reasonable, then the recovery time t_r under the nominal case can be approximated as the following formula:

$$\begin{aligned} t_r &\approx L + t_m + 2\lambda e \\ &= \left(1 + \beta + 2\beta e + \frac{1}{\beta + 1}\right) L_m. \end{aligned}$$

Many simulations show that a more accurate and simpler formula for the 5% error band is

$$t_r \approx (7\beta + 2) L_m.$$

Hence, from the recovery-time point of view, it is better to choose $\beta = 0 \sim 1$, *i.e.*, $\lambda = (0 \sim 1) L_m$.

For a given control parameter $\beta = \lambda / L_m$, the achievable specifications of the step-disturbance response are shown in Figure 11.4.

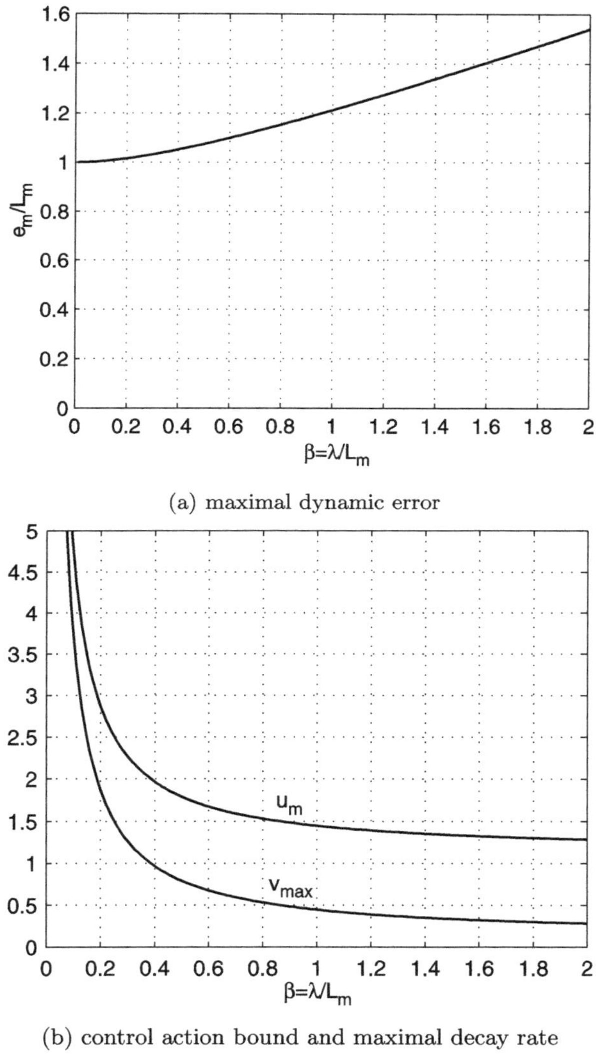

Fig. 11.4 Achievable specifications of the disturbance response

11.4 Robust Stability Regions

Three cases of different uncertainties (gain uncertainties, dead-time uncertainties, and both) will be considered below.

The complementary sensitivity function of the nominal system is

$$T(s) = Q(s)e^{-L_m s}.$$

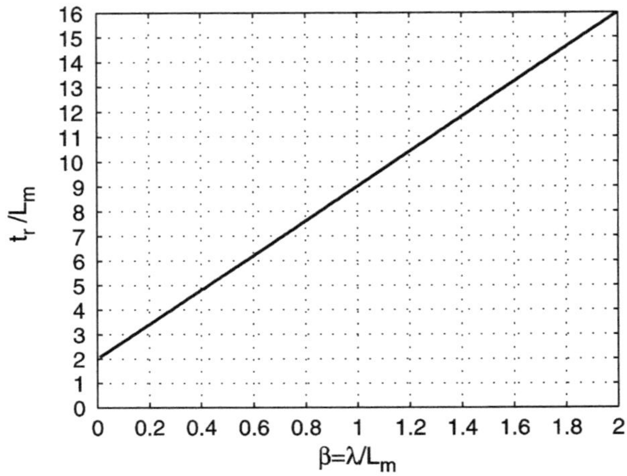

(c) approximate recovery time

Fig. 11.4 (Continued)

According to the well-known small-gain theorem, the closed-loop system is robustly stable if $\|Q(s)\|_\infty < \frac{1}{\|\Delta(s)\|_\infty}$ for a multiplicative uncertainty $\Delta(s) \in H_\infty$. Since the steady-state gain of $Q(s)$ is 1, which means there exists an integral effect in the controller, a multiplicative uncertainty does not affect the steady-state performance of the system.

11.4.1 With Gain Uncertainties

The magnitude response of $Q(s)$ is

$$\left|Q(j\omega)\right| = \frac{\sqrt{1+(2\lambda+L_m)^2\omega^2}}{1+\lambda^2\omega^2},$$

and the peak value is

$$\left|Q(j\omega)\right|_{\max} = \frac{(2\beta+1)^2}{2\beta\sqrt{(\beta+1)(3\beta+1)}}.$$

Hence, the following theorem holds:

Theorem 11.4 *For a gain uncertainty* $\Delta(s) = \frac{\Delta K}{K}$, *the closed-loop system is robustly stable if* $\frac{|\Delta K|}{K} < \frac{2\beta\sqrt{(\beta+1)(3\beta+1)}}{(2\beta+1)^2}$.

As $\beta \to \infty$, $\frac{2\beta\sqrt{(\beta+1)(3\beta+1)}}{(2\beta+1)^2} \to \frac{\sqrt{3}}{2}$. This provides the bound $\frac{\sqrt{3}}{2}$ for the gain uncertainty to guarantee the robust stability. This bound is about 0.63 when $\beta = 1$.

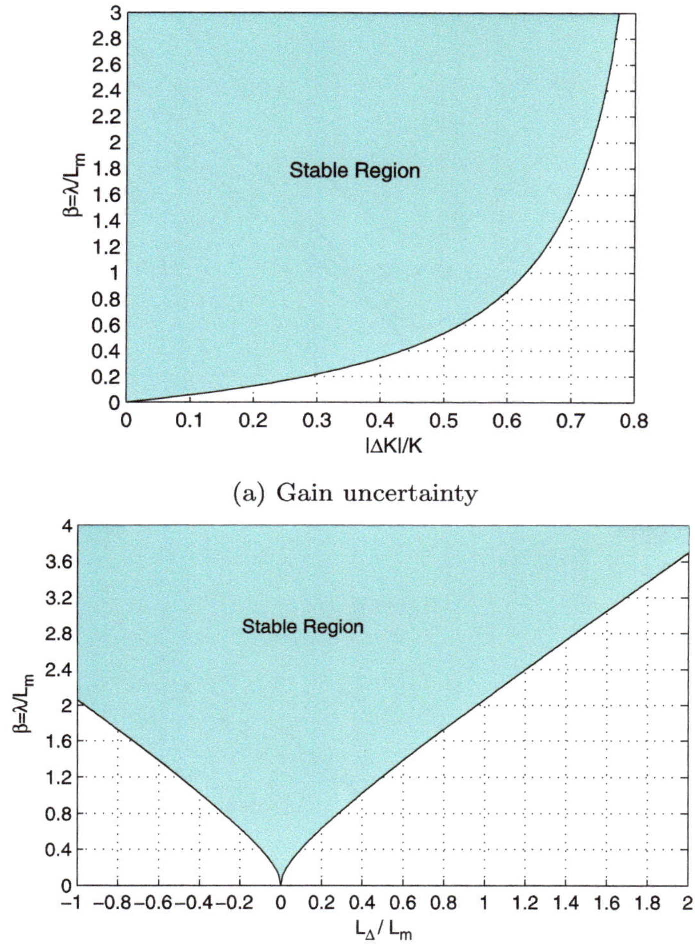

(a) Gain uncertainty

(b) Dead-time uncertainty

Fig. 11.5 Control parameter *vs.* uncertainty

The relationship between the control parameter and the allowable gain uncertainty is shown in Figure 11.5(a).

11.4.2 With Dead-time Uncertainties

Assume there exists a dead-time uncertainty, *i.e.*,

$$G_p(s) = \frac{K}{s} \quad \text{and} \quad L = L_m + L_\Delta \quad (-L_m \leq L_\Delta).$$

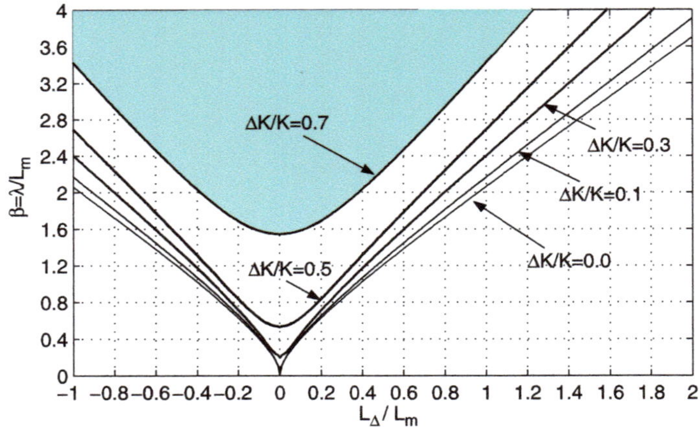

(c) Positive gain uncertainty with dead-time uncertainty

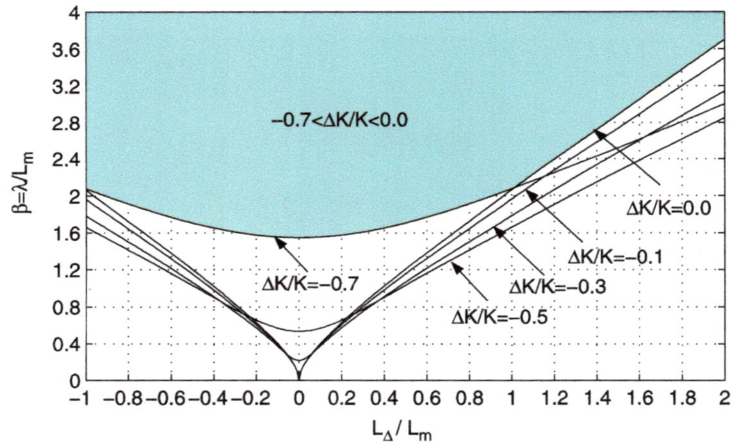

(d) Negative gain uncertainty with dead-time uncertainty

Fig. 11.5 (Continued)

This uncertainty can be converted to a multiplicative uncertainty $\Delta(s) = e^{-L_\Delta s} - 1$. Hence, the robust stability is guaranteed if the following condition is met:

$$\begin{aligned}|Q(j\omega)| &= \frac{\sqrt{1+(2\lambda+L_m)^2\omega^2}}{1+\lambda^2\omega^2}\\ &< \frac{1}{|e^{-jL_\Delta\omega}-1|} = \frac{1}{\sqrt{2-2\cos(L_\Delta\omega)}} \quad (\forall\omega>0),\end{aligned}$$

that is,

$$2-2\cos(L_\Delta\omega) < \frac{(1+\lambda^2\omega^2)^2}{1+(2\lambda+L_m)^2\omega^2}$$

$$= \frac{\left(1+\left(\frac{\lambda}{L_\Delta}\right)^2 (L_\Delta\omega)^2\right)^2}{1+\left(\frac{2\lambda+L_m}{L_\Delta}\right)^2 (L_\Delta\omega)^2}$$

$$= \frac{\left(1+\left(\frac{\beta}{L_\Delta/L_m}\right)^2 (L_\Delta\omega)^2\right)^2}{1+\left(\frac{2\beta+1}{L_\Delta/L_m}\right)^2 (L_\Delta\omega)^2} \quad (\forall\omega>0).$$

There does not exist an analytical solution for β. Denote $\omega_\Delta = L_\Delta\omega$, $f_1(\omega_\Delta) = 2-2\cos(\omega_\Delta)$, and $f_2(\omega_\Delta) = \frac{(1+(\frac{\beta}{L_\Delta/L_m})^2\omega_\Delta^2)^2}{1+(\frac{2\beta+1}{L_\Delta/L_m})^2\omega_\Delta^2}$. Since $f_1(\omega_\Delta)$ and $f_2(\omega_\Delta)$ are all even functions with respect to ω_Δ, only the case with $\omega_\Delta \geq 0$ is considered. The only intersection of $f_1(\omega_\Delta)$ and $f_2(\omega_\Delta)$ can be obtained when

$$f_1(\omega_\Delta) = f_2(\omega_\Delta) \quad \text{and} \quad \dot{f}_1(\omega_\Delta) = \dot{f}_2(\omega_\Delta).$$

This can be solved numerically. The solutions of the lower bound of β to guarantee the robust stability for $\frac{L_\Delta}{L_m} \in [-1,\ 2]$ are shown in Figure 11.5(b). For a large dead-time uncertainty $\frac{L_\Delta}{L_m} > 1$, the lower bound of the control parameter is about $1.625\frac{L_\Delta}{L_m} + 0.45$.

11.4.3 With Dead-time and Gain Uncertainties

Assume that there exist uncertainties in both dead time L and gain k, *i.e.*,

$$G_p(s) = \frac{K}{s}\left(1+\frac{\Delta K}{K}\right) \quad \text{and} \quad L = L_m + L_\Delta \quad (-L_m \leq L_\Delta).$$

Then the corresponding multiplicative uncertainty is

$$\Delta(s) = \left(1+\frac{\Delta K}{K}\right)e^{-L_\Delta s} - 1.$$

The robust stability is guaranteed if the following condition is met:

$$\frac{\sqrt{1+(2\lambda+L_m)^2\omega^2}}{1+\lambda^2\omega^2} < \frac{1}{\sqrt{1+\left(1+\frac{\Delta K}{K}\right)^2 - 2\left(1+\frac{\Delta K}{K}\right)\cos(L_\Delta\omega)}} \quad (\forall\omega>0).$$

This inequality cannot be analytically solved, either. For some given gain uncertainties $\frac{\Delta K}{K}$, the relationship between the control parameter β and the dead-time uncertainty can be obtained numerically, as shown in Figure 11.5(c, d). For a positive gain uncertainty together with a dead-time uncertainty, the stability region is sharp. The larger the gain uncertainty, the sharper the stability region. For a negative gain uncertainty together with a dead-time uncertainty, the stability region is blunt. The larger the (absolute) gain uncertainty, the blunter the stability region. For a gain uncertainty $q \leq \frac{\Delta K}{K} \leq p$ ($p, q \in [-\frac{\sqrt{3}}{2}, \frac{\sqrt{3}}{2}]$), the stability region is the intersection of the corresponding stability regions for $\frac{\Delta K}{K} = p$ and $\frac{\Delta K}{K} = q$. For example, the shaded area in Figure 11.5(d) is the stability region for a gain uncertainty

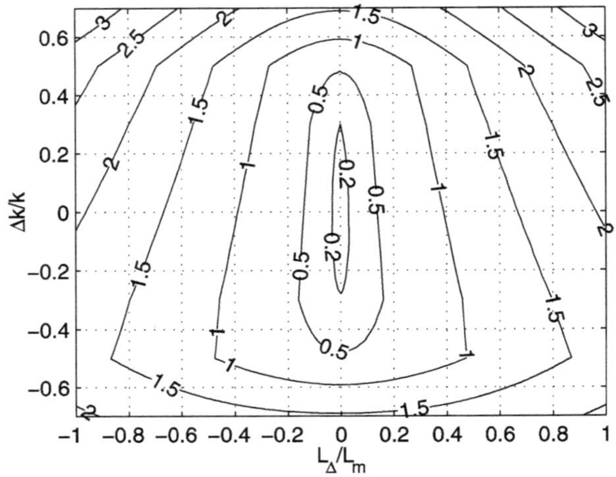

Fig. 11.6 Rough contour of β *vs*. uncertainties

$-0.7 \le \frac{\Delta K}{K} \le 0.0$, while the shaded area in Figure 11.5(c) is the stability region for a gain uncertainty $-0.7 \le \frac{\Delta K}{K} \le 0.7$ because the stability region for $\frac{\Delta K}{K} = 0.7$ is included by that of $\frac{\Delta K}{K} = -0.7$. In order to obtain a better view of the relationship, a rough contour of β with respect to $\frac{\Delta K}{K}$ and $\frac{L_\Delta}{L_m}$ is shown in Figure 11.6. This is very useful for determining β to meet the uncertainties. Figure 11.6 indicates that the nominal gain should be chosen near the maximal possible gain rather than the centre of the gain range.

It should be pointed out that the stability regions obtained above are sufficient conditions but not necessary conditions.

11.5 Stability of the Controller

It is always advantageous to have a stable controller, in addition to a stable closed-loop control system. The stability of the controller itself is analysed in this section.

The characteristic equation of the controller can be equivalent to

$$\frac{(\lambda s+1)^2}{(2\lambda+L)s+1} = e^{-Ls}. \tag{11.15}$$

The stability can be determined using the dual-locus diagram method [114, 157]. The loci of the both sides in (11.15) are shown in Figure 11.7. If the two loci do not intersect, then the controller is stable; otherwise it is unstable. In other words, if the locus of $\frac{(\lambda s+1)^2}{(2\lambda+L)s+1}$ crosses the unit circle at which before e^{-Ls} locus arrives, then the system is stable; if e^{-Ls} arrives before $\frac{(\lambda s+1)^2}{(2\lambda+L)s+1}$, then the controller is unstable.

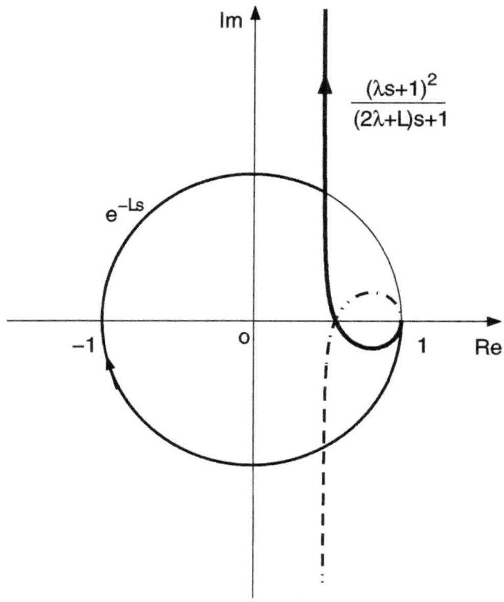

Fig. 11.7 The dual-locus diagram of the controller

The locus of $\frac{(\lambda s+1)^2}{(2\lambda+L)s+1}$ intersects with the unit circle at frequency $\omega_c = \frac{\sqrt{2\beta^2+4\beta+1}}{\beta^2 L}$ ($\beta = \frac{\lambda}{L}$). The phase angle of $\frac{(\lambda s+1)^2}{(2\lambda+L)s+1}$ at ω_c is

$$\phi_\lambda = 2\arctan\frac{\sqrt{2\beta^2+4\beta+1}}{\beta} - \arctan\frac{(2\beta+1)\sqrt{2\beta^2+4\beta+1}}{\beta^2}.$$

If the locus of e^{-Ls} also arrives this point at ω_c, then the following equation should be met:

$$\phi_\lambda = 2\pi - L\omega_c,$$

i.e.,

$$2\arctan\frac{\sqrt{2\beta^2+4\beta+1}}{\beta} - \arctan\frac{(2\beta+1)\sqrt{2\beta^2+4\beta+1}}{\beta^2} = 2\pi - \frac{\sqrt{2\beta^2+4\beta+1}}{\beta^2}.$$

This equation has a unique solution at about 0.63. For $\beta > 0.63$, the locus of $\frac{(\lambda s+1)^2}{(2\lambda+L)s+1}$ crosses the unit circle before e^{-Ls} locus arrives at the same point. Hence, the controller is stable for $\beta > 0.63$.

11.6 Conclusions

Many different control schemes for integral processes with dead time have resulted in the same disturbance response, which can be divided into three stages. The first two stages are not controllable, and the third stage is determined by the control parameter. As a result, the minimum of the normalised maximal dynamic error is not less than L_m. Four specifications: the (normalised) maximal dynamic error, the (normalised) control action bound, the maximal decay rate, and an approximate recovery time are used to characterise the disturbance response. For a given (reasonable) maximal dynamic error, the control parameter is determined in terms of the *Lambert W* function. For a chosen control parameter, there exists a requirement of the control action bound. Moreover, the (sufficient) stability regions for gain uncertainties, dead-time uncertainties, and both of them are shown in figures. A sufficient condition on the (relative) gain-uncertainty bound is $\frac{\sqrt{3}}{2}$. The controller itself is stable, provided that $\beta > 0.63$. Hence, these lead to the quantitative design of the disturbance-observer-based control scheme for integral processes according to the requirements of the disturbance response, the robustness, and the stability of the controller itself.

Chapter 12
Practical Issues

Various control schemes for integral processes with dead time are discussed in the previous chapters. Some practical issues concerned with the controller implementation will be discussed in this chapter and then verified with experiments carried out on a laboratory-scale setup.

12.1 The Control Scheme Under Consideration

The integral processes with dead time can be described as

$$G(s) = G_p(s)e^{-\tau s} = \frac{K}{s}e^{-\tau s}, \tag{12.1}$$

where the pure dead time τ and the static gain K are both positive. The disturbance-observer-based control scheme presented in Chapter 10 is taken as an example to facilitate the discussions in the sequel. Similar issues exist with other control schemes for integral processes with dead time. The scheme is revisited in Figure 12.1(a) for the readers' convenience, where τ_m is the estimated dead time, and $G_m(s)$ is the nominal delay-free part. The controller $C(s) = \frac{1}{KT}$ is designed to be proportional, and the low-pass filter

$$Q(s) = \frac{(2\lambda + \tau_m)s + 1}{(\lambda s + 1)^2} \tag{12.2}$$

is designed to reject step disturbances with no static error, where λ is a design parameter to compromise the disturbance rejection, robustness and, possibly, the control action bound [163].

Under the nominal condition, $G_m(s) = G_p(s) = \frac{K}{s}$, and $\tau_m = \tau$. The set-point response and disturbance response are respectively

$$G_{yr}(s) = \frac{1}{Ts+1}e^{-\tau_m s}$$

and

$$G_{yd}(s) = \frac{K}{s}\left(1 - Q(s)e^{-\tau_m s}\right)e^{-\tau_m s}. \tag{12.3}$$

A. Visioli, Q.-C. Zhong, *Control of Integral Processes with Dead Time*,
Advances in Industrial Control,
DOI 10.1007/978-0-85729-070-0_12, © Springer-Verlag London Limited 2011

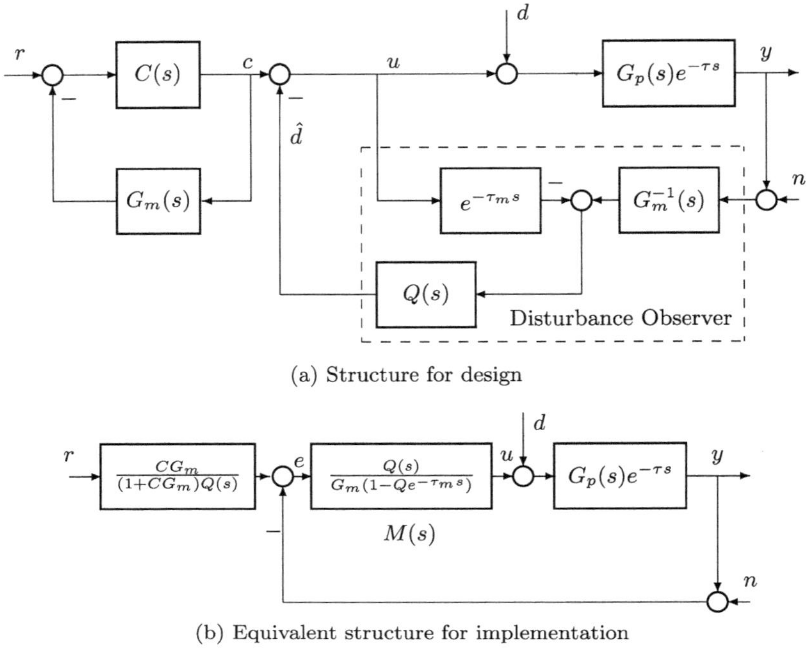

Fig. 12.1 Disturbance observer-based control scheme

The scheme shown in Figure 12.1(a) is not internally stable nor causal. An algebraically equivalent but causal and internally stable structure is shown in Figure 12.1(b), as proposed in [162]. The implementation of the block

$$M(s) = \frac{Q(s)}{G_m(s)(1 - Qe^{-\tau_m s})}$$

needs to be very careful. Otherwise, problems may occur.

12.2 Zero Static Error

The block diagram shown in Figure 12.2(a) for $M(s)$ is implementable because the block $\frac{Q(s)}{G_m(s)}$ is causal. However, if the block $M(s)$ is implemented in this way, then the system is not able to reject step disturbances completely. There is a non-zero static error because the block

$$\frac{Q(s)}{G_m(s)} = \frac{(2\lambda + \tau_m)s^2 + s}{K(\lambda s + 1)^2}$$

contains a differentiator and any constant error e is ignored by the controller. This causes a non-zero static error. As a matter of fact, since the static gain of $Q(s)$ is unity, the local positive feedback loop in Figure 12.2(a) contains an integral action.

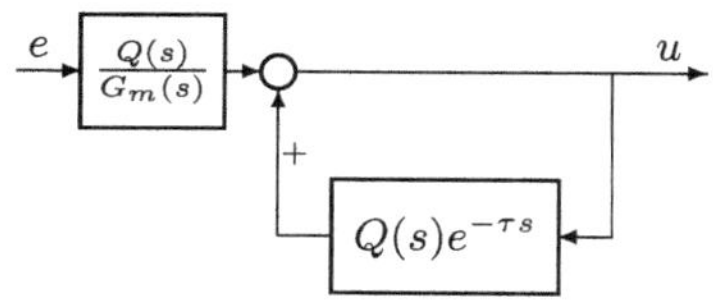

(a) Non-zero static error

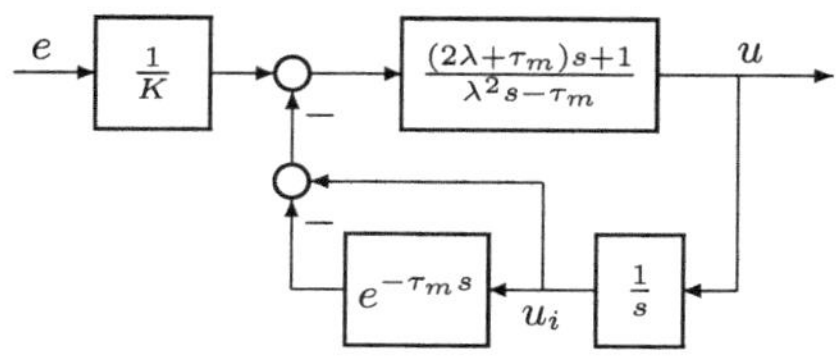

(b) Internal instability

Fig. 12.2 Wrong implementation schemes for $M(s)$

The differential action in the previous block cancels this integral action. This should be avoided in practical implementation.

12.3 Internal Stability

One special property of the control plant is the instability due to the pole at the origin. This makes the structure shown in Figure 12.1(a) internally unstable because of the pole-zero cancellation at $s=0$ between the block $G_m^{-1}(s)$, the inverse of the delay-free model, and the plant $G_p(s)e^{-\tau s}$. The internal stability problem can be solved if a discretised version of the controller is implemented [81, 82]. For an analog controller, the equivalent structure for implementation shown in Figure 12.1(b) guarantees the internal stability if implemented properly. As discussed in the previous subsection, the implementation of $M(s)$ shown in Figure 12.2(a) causes non-zero static errors when the system is subject to step a disturbance. The cancellation between the differential action in $\frac{Q(s)}{G_m(s)}$ and the integral action in the local loop should be removed in implementation. The block $M(s)$ can be algebraically equivalent to

$$\begin{aligned}
M(s) &= \frac{Q(s)}{G_m(s)(1-Qe^{-\tau_m s})} \\
&= \frac{1}{K}\frac{(2\lambda+\tau_m)s+1}{\lambda^2 s-\tau_m+((2\lambda+\tau_m)s+1)\frac{1-e^{-\tau_m s}}{s}} \\
&= \frac{1}{K}\frac{\frac{(2\lambda+\tau_m)s+1}{\lambda^2 s-\tau_m}}{1+\frac{(2\lambda+\tau_m)s+1}{\lambda^2 s-\tau_m}\cdot\frac{1-e^{-\tau_m s}}{s}}.
\end{aligned}$$

This can be described as shown in Figure 12.2(b) as a local feedback loop. The signal u_i in Figure 12.2(b), *i.e.*, the state of the integrator, is unbounded, and hence

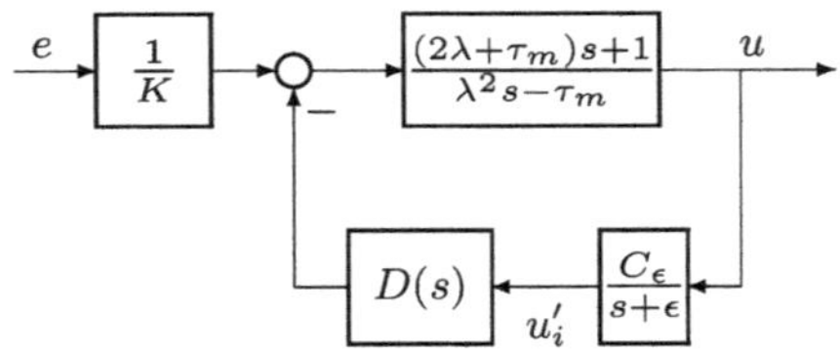

Fig. 12.3 Correct implementation scheme of $M(s)$

the system is not internally stable even when the output of the whole system is stable because the block $\frac{1-e^{-\tau_m s}}{s}$ is split into the cascade of an integrator and $1 - e^{-\tau_m s}$, which causes a zero-pole cancellation at $s = 0$. One way to avoid this is to use the approach proposed in [158, 160] to approximate $\frac{1-e^{-\tau_m s}}{s}$ as

$$\frac{1 - e^{-\tau_m s}}{s} \approx \frac{C_\epsilon}{s + \epsilon} \cdot D(s),$$

with

$$C_\epsilon = \frac{\frac{\tau_m}{N}\epsilon}{1 - e^{-\epsilon \tau_m / N}},$$

and

$$D(s) = \left(1 - e^{-\frac{\tau_m}{N}(s+\epsilon)}\right) \cdot \sum_{i=0}^{N-1} e^{-i\frac{\tau_m}{N}s},$$

where $\epsilon > 0$ is a small enough number, and N is a natural number. The longer the delay, the bigger the N is needed. Then, the block $M(s)$ can be implemented as shown in Figure 12.3. With this implementation, the closed-loop system is internally stable. Moreover, there is no static error. The only downside is that the performance will be slightly worse than the designed performance due to the approximation. This can be improved by choosing a small ϵ and a big N.

12.4 Experimental Results

12.4.1 The Experimental Setup

The laboratory experimental setup employed to verify the findings in the previous sections is shown in Figure 12.4. This is a small perspex tower-type tank with a section area of $A = 40\,\text{cm}^2$, of which the water level is regulated by a PC-based controller. The tank is filled with water by means of a pump whose speed is controlled by a DC voltage, in the range 0–5 V, through a pulse-width-modulation (PWM) circuit. Then, an outlet at the base allows the water to return to the reservoir. The measure of the level of the water is given by a capacitive-type probe that provides

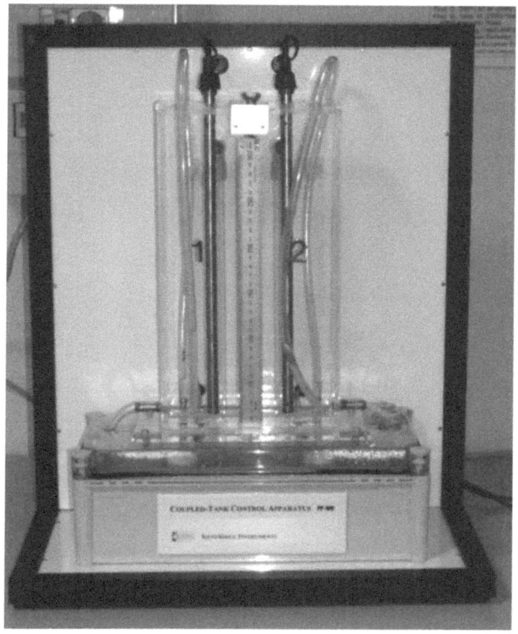

Fig. 12.4 The experimental setup (only one tank has been employed in the experiments)

an output signal between 0 (empty tank) and 5 V (full tank). The process can be modelled by the following differential equation:

$$\frac{dy(t)}{dt} = \frac{1}{A}\big(q_i(t) - q_o(t)\big),$$

where y is the water level (*i.e.*, the process variable), q_i is the input flow rate, and q_o is the output flow rate. The input flow rate q_i is linearly dependent on the voltage applied to the DC pump. The output flow rate can be calculated by measuring the level $y(t)$ because $q_o(t) = \alpha\sqrt{2gy(t)}$, where g is the gravitational acceleration constant, and α is the cross-sectional area of the output orifice. This is then added to the control variable u via a local feedback loop to obtain the input flow rate q_i, i.e., $u(t) + q_0(t) = q_i(t)$, which is then used to determine the speed of the DC motor (pump). Hence, the control variable u is

$$u(t) = q_i(t) - q_o(t).$$

In this way an integral process is obtained as

$$\frac{dy(t)}{dt} = \frac{1}{A}u(t).$$

Dead time also occurs in the process because of the length of the pipe. The transfer function of the process, from the control variable u to the voltage corresponding to

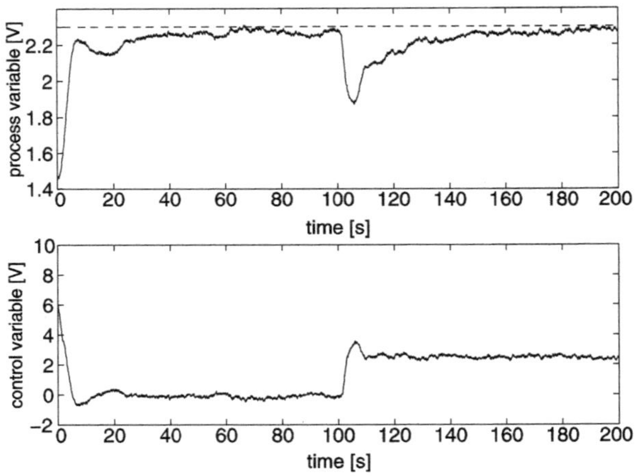

Fig. 12.5 Experimental results with the control scheme of Figure 12.2(a)

the water level, has been estimated as follows by means of a simple open-loop step response experiment,

$$G(s) = \frac{0.026}{s} e^{-1.5s}.$$

12.4.2 The Scheme Shown in Figure 12.2(a)

The control parameters are chosen as $T = 5$ and $\lambda = 8$. A set-point change from 1.5 V to 2.3 V has been applied to the control system and, after 100 s, a disturbance $d = -2.5$ V has been applied to the control variable. The resulting process variable, together with the set-point signal, is shown in Figure 12.5. As expected, a non-zero static error appears.

12.4.3 The Scheme Shown in Figure 12.2(b)

The same experiment as described in the previous subsection has been repeated for the scheme shown in Figure 12.2(b). The resulting process and control variables are plotted in Figure 12.6(a). The static error has disappeared, but, as expected, the system is internally unstable as can be seen from the plot of the variable u_i (*i.e.*, the state of the integrator) shown in Figure 12.6(b). This variable becomes unbounded eventually.

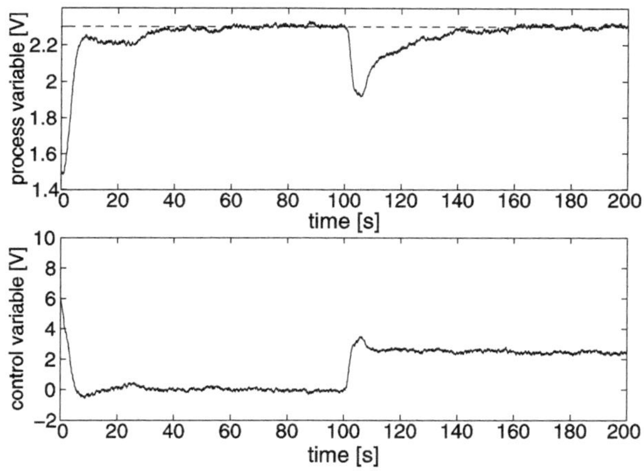

(a) Process variable and control variable

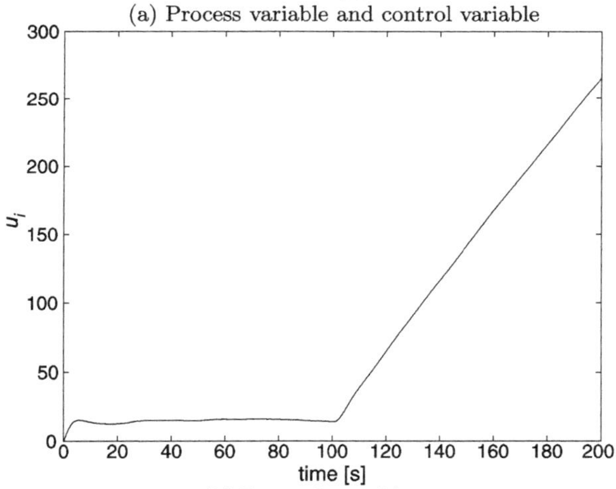

(b) Internal variable u_i

Fig. 12.6 Experimental results with the control scheme of Figure 12.2(b)

12.4.4 The Scheme Shown in Figure 12.3

Then, the scheme of Figure 12.3 has been employed for the same control task. The control parameters are selected as $T = 5$, $\lambda = 8$, $\epsilon = 0.01$, and $N = 1$. Results are shown in Figure 12.7(a). It appears that the control scheme is effective in both the set-point response and disturbance rejection. In order to evaluate the physical meaning of the parameter λ, the same experiment as before has been repeated with different values of $\lambda = 4$ and $\lambda = 12$. The resulting process and control variables are plotted in Figures 12.7(b) and 12.7(c), respectively. The control system is more aggressive when the value of λ is decreased, and it is more sluggish when λ is increased, as expected.

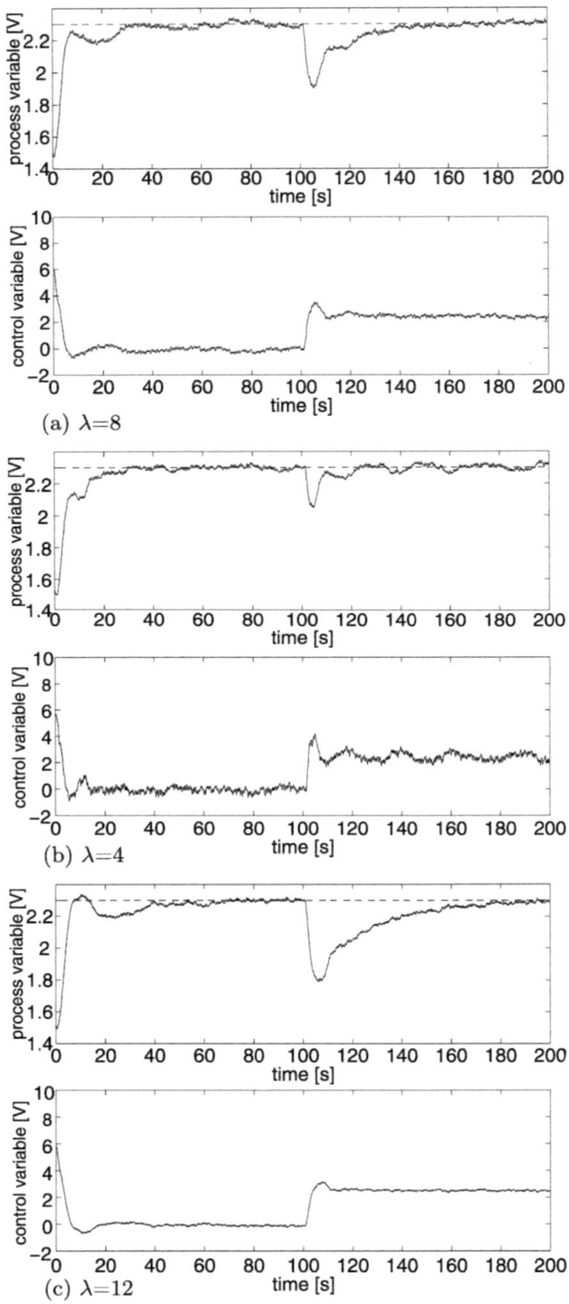

Fig. 12.7 Experimental results with the control scheme of Figure 12.3

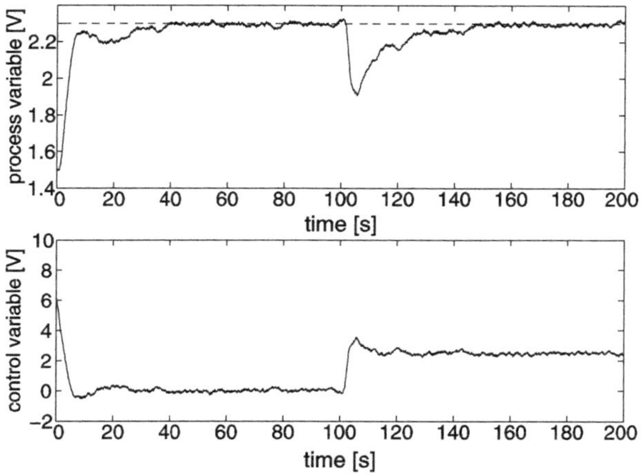

Fig. 12.8 Experimental results for a digital controller ($\lambda = 8$)

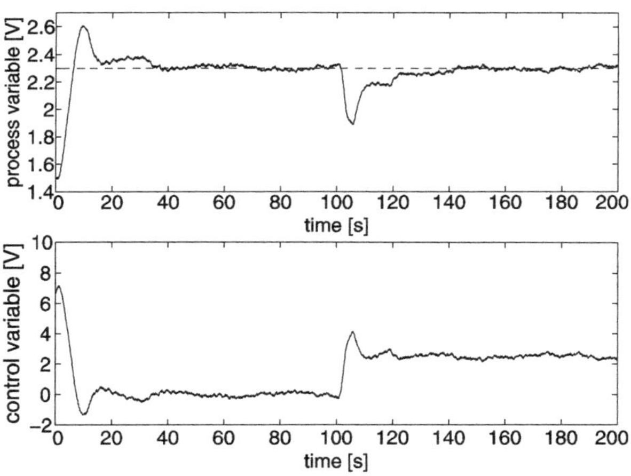

Fig. 12.9 Experimental results with a PI controller

For the sake of comparison, a digital controller has also been implemented. A transfer function $M(z)$ can be obtained straightforward after having discretised $Q(s)$, $G_m(s)$, and $e^{-\tau_m s}$ by applying the backward difference approach. Then, the obtained $M(z)$ is

$$M(z) = \frac{10.5168z^{15}(z - 0.9943)}{N(z)}$$

with

$$N(z) = (z-1)(z+0.7582)(z^2 - 1.649z + 0.7333)(z^2 + 1.394z + 0.5764)$$
$$\times (z^2 - 1.249z + 0.6583)(z^2 + 1.047z + 0.5811)(z^2 - 0.7115z + 0.6238)$$

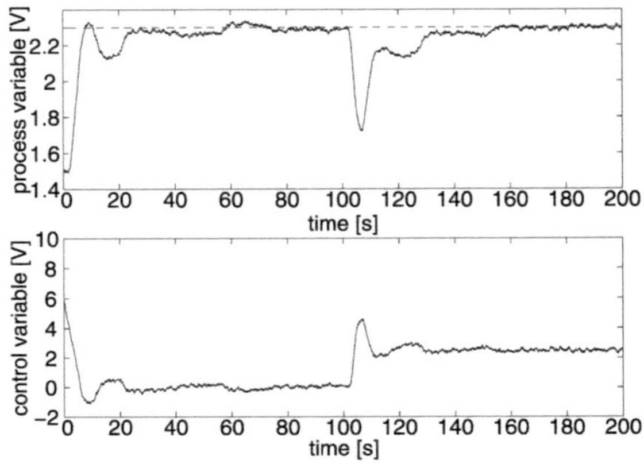

(a) With the control scheme of Fig. 12.3 (λ=8)

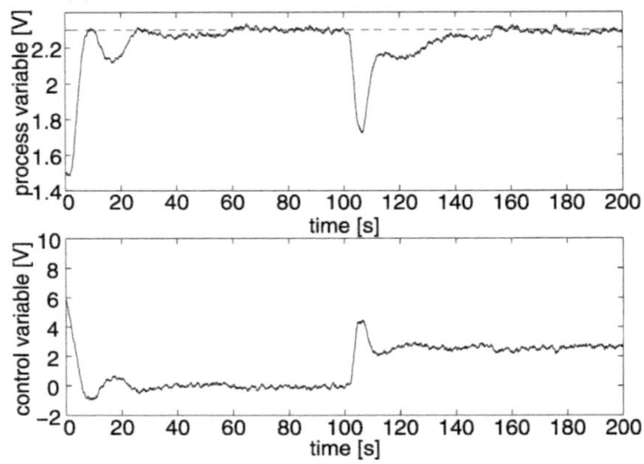

(b) With the control scheme of Fig. 12.3 (λ=12)

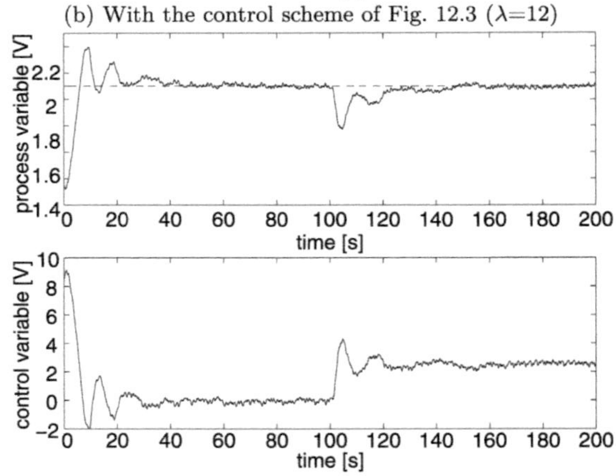

(c) With PI control scheme ($K_p = 8.48$, $T_i = 15$)

Fig. 12.10 Experimental results for robustness

$$\times (z^2 + 0.5255z + 0.5896)(z^2 - 0.09093z + 0.603),$$

which was employed in the scheme of Figure 12.1(b) after sampling the control error e at the rate of 0.1 s (note that the resulting controller is of the 16th order). Results related to the case of $\lambda = 8$ are shown in Figure 12.8. It can be seen that there is no significant difference between the use of an analog or digital controller.

12.4.5 Comparison with a PI Controller

In order to evaluate better the performance of the disturbance observer-based control scheme, comparison with a standard PI controller in the unity-feedback control scheme has been performed. The controller is

$$C_{\mathrm{PI}}(s) = K_p\left(1 + \frac{1}{T_i s}\right),$$

where the proportional gain K_p and the integral time constant T_i have been selected by applying the tuning rule proposed in [15], *i.e.*, $K_p = \frac{0.3306}{KL} = 8.48$, $T_i = 10L = 15$. The experimental results are shown in Figure 12.9. The disturbance rejection is very good, but the set-point response has a large overshoot. This clearly shows the benefit of the two-degree-of-freedom control scheme. It is able to provide good performance for both the set-point response and disturbance rejection. Obviously, being of a single degree-of-freedom, the PI controller is able to obtain satisfactory performance in disturbance rejection at the expense of a significant overshoot in the set-point response. In general, it is not easy to select the most appropriate tuning rule to meet the requirement for both.

12.4.6 Robustness with Respect to Changes in the Dead Time

Experiments have also been performed on the same system as before but with an additional input dead time of 1 s added via software delay. The control scheme of Figure 12.3 with $\lambda = 8$ and $\lambda = 12$ is compared with the PI control scheme described in the previous subsection. The results are shown in Figure 12.10.

12.5 Conclusions

In this chapter, some practical issues in controller implementation for integral processes with dead time are discussed and verified with experimental results. It is shown that with proper implementation, these problems can be eliminated.

References

1. Ang, K.H., Chong, G., Li, Y.: PID control systems analysis, design, and technology. IEEE Trans. Control Syst. Technol. **13**, 559–576 (2005)
2. Arbogast, J.E., Cooper, D.J.: Extension of IMC tuning correlations for non-self regulating (integrating) processes. ISA Trans. **46**, 303–311 (2007)
3. Åström, K.J., Hägglund, T.: Automatic tuning of simple controllers with specification on phase and amplitude margins. Automatica **20**(5), 655–651 (1984)
4. Åström, K.J., Hägglund, T.: Automatic Tuning of PID Controllers. Instrument Society of America, Research Triangle Park (1988)
5. Åström, K.J., Hägglund, T.: PID Controllers: Theory, Design, and Tuning, 2nd edn. Instrument Society of America, Research Triangle Park (1995)
6. Åström, K.J., Hägglund, T.: Advanced PID Control. Instrument Society of America, Research Triangle Park (2006)
7. Åström, K.J., Wittenmark, B.: Computer-Controlled Systems—Theory and Design. Prentice Hall, Upper Saddle River (1997)
8. Åström, K.J., Hägglund, T., Hang, C.C., Ho, W.K.: Automatic tuning and adaptation for PID controllers—a survey. Control Eng. Pract. **1**, 699–714 (1993)
9. Åström, K.J., Hang, C.C., Lim, B.C.: A new Smith predictor for controlling a process with an integrator and long dead-time. IEEE Trans. Autom. Control **39**, 343–345 (1994)
10. Bellman, R., Cooke, K.L.: Differential-Difference Equations. Academic Press, London (1963)
11. Beschi, M., Piazzi, A., Visioli, A.: On the practical implementation of a noncausal feedforward technique for PID control. In: Proceedings European Control Conference, pp. 1806–1811, Budapest, HU, 2009
12. Bodson, M.: Adaptive algorithm for the tuning of two input shaping methods. Automatica **34**, 771–776 (1998)
13. Camacho, O., De la Cruz, F.: Smith predictor based-sliding mode controller for integrating processes with elevated deadtime. ISA Trans. **43**, 257–270 (2004)
14. Chidambaram, M., Padma Sree, R.: A simple method of tuning PID controllers for integrator/dead-time processes. Comput. Chem. Eng. **27**, 211–215 (2003)
15. Chien, I.L., Fruehauf, P.S.: Consider IMC tuning to improve performance. Chem. Eng. Prog. **86**(10), 33–41 (1990)
16. Chien, I.-L., Peng, S.C., Liu, J.H.: Simple control method for integrating processes with long deadtime. J. Process Control **12**, 391–404 (2002)
17. Corless, R.M., Gonnet, G.H., Hare, D.E.G., Jeffrey, D.J., Knuth, D.E.: On the Lambert W function. Adv. Comput. Math. **5**, 329–359 (1996). Available at http://www.apmaths.uwo.ca/~rcorless/frames/papers.htm, accessed on 3/9/2005
18. Datta, A., Ho, M.-T., Bhattacharyya, S.P.: Structure and Synthesis of PID Controllers. Springer, London (2000)

A. Visioli, Q.-C. Zhong, *Control of Integral Processes with Dead Time*, Advances in Industrial Control,
DOI 10.1007/978-0-85729-070-0,

© Springer-Verlag London Limited 2011

19. Dejonckheere, J., Disney, S.M., Lambrecht, M.R., Towill, D.R.: Measuring and avoiding the bullwhip effect: A control theoretic approach. Eur. J. Oper. Res. **147**(3), 567–590 (2003)
20. Devasia, S., Chen, D., Paden, B.: Nonlinear inversion-based output tracking. IEEE Trans. Autom. Control **41**, 930–943 (1996)
21. Disney, S.M., Towill, D.R., Warburton, R.D.H.: On the equivalence of the control theoretic, differential and difference equation approaches to solving the production and inventory control system design problem. Int. J. Prod. Econ. **101**(1), 194–208 (2006)
22. Doyle, J.C., Francis, B.A., Tannenbaum, A.R.: Feedback Control Theory. Macmillan, New York (1992)
23. Dym, H., Georgiou, T., Smith, M.C.: Explicit formulas for optimally robust controllers for delay systems. IEEE Trans. Autom. Control **40**(4), 656–669 (1995)
24. Endo, S., Kobayashi, H., Kempf, C.J., Kobayashi, S., Tomizuka, M., Hori, Y.: Robust digital tracking controller design for high-speed positioning systems. Control Eng. Pract. **4**(4), 527–536 (1996)
25. Eriksson, P.-G., Isaksson, A.J.: Some aspects of control loop performance monitoring. In: Proceedings IEEE International Conference on Control Applications, pp. 1029–1034, Glasgow, UK, 1994
26. Friman, M., Waller, K.V.: Autotuning of multiloop control systems. Ind. Eng. Chem. Res. **33**, 1708–1717 (1994)
27. Gémez-Stern, F., Fornés, J.M., Rubio, F.R.: Dead-time compensation for ABR traffic control over ATM networks. Control Eng. Pract. **10**(5), 481–491 (2002)
28. Guanghui, Z., Huihe, S.: A simple anti-windup compensation for modified Smith predictor. In: Proceedings American Control Conference, pp. 4859–4863, Minneapolis, Minnesota, 2006
29. Guanghui, Z., Huihe, S.: Anti-windup design for the design for the controllers of integrating processes with long delay. J. Syst. Eng. Electron. **18**(2), 297–303 (2007)
30. Guanghui, Z., Feng, Q., Huihe, S.: Robust tuning method for modified Smith predictor. J. Syst. Eng. Electron. **18**(1), 89–94 (2007)
31. Hägglund, T.: Automatic detection of sluggish control loops. Control Eng. Pract. **7**, 1505–1511 (1999)
32. Hägglund, T., Åström, K.J.: Supervision of adaptive control algorithms. Automatica **36**(2), 1171–1180 (2000)
33. Hara, S., Yamamoto, Y., Omata, T., Nakano, M.: Repetitive control system: a new type servo system for periodic exogenous signals. IEEE Trans. Autom. Control **33**, 659–668 (1988)
34. Harris, T.J.: Assessment of control loop performance. Can. J. Chem. Eng. **67**, 856–861 (1989)
35. Ho, W.K., Feng, E.B., Gan, O.P.: A novel relay auto-tuning technique for processes with integration. Control Eng. Pract. **4**(7), 923–928 (1996)
36. Hong, K., Nam, K.: A load torque compensation scheme under the speed measurement delay. IEEE Trans. Ind. Electron. **45**(2), 283–290 (1998)
37. Howard, R., Cooper, D.J.: Performance assessment of non-self regulating controllers in a cogeneration power plant. Appl. Energy **86**, 2121–2129 (2009)
38. Huang, B., Shah, S.L.: Performance Assessment of Control Loop. Springer, London (1999)
39. Huang, H.-P., Jeng, J.-C.: Monitoring and assessment of control performance for single loop systems. Ind. Eng. Chem. Res. **41**, 1297–1309 (2002)
40. Huang, K., Shan, L., Zhu, Q., Qian, J.: A totally heat-integrated distillation column (THIDiC)—the effect of feed pre-heating by distillate. Appl. Therm. Eng. **28**, 856–864 (2008)
41. Hunt, L.R., Meyer, G.: Stable inversion for nonlinear systems. Automatica **33**(8), 1549–1554 (1997)
42. Hunt, L.R., Meyer, G., Su, R.: Noncausal inverses for linear systems. IEEE Trans. Autom. Control **41**, 608–611 (1996)
43. Ingimundarson, A., Hägglund, T.: Robust tuning procedures of dead-time compensating controllers. Control Eng. Pract. **9**, 1195–1208 (2001)

44. Izmailov, R.: Analysis and optimization of feedback control algorithms for data transfers in high-speed networks. SIAM J. Control Optim. **34**(5), 1767–1780 (1996)
45. Jelali, M.: An overview of control performance assessment technology and industrial applications. Control Eng. Pract. **14**, 441–466 (2006)
46. Kaya, I.: Controller design for integrating processes using user-specified gain and phase margin specifications and two degree-of-freedom IMC structure. In: Proceedings IEEE International Conference on Control Applications, pp. 898–902, Istanbul, Turkey, 2003
47. Kaya, I.: A PI–PD controller design for control of unstable and integrating processes. ISA Trans. **42**, 111–121 (2003)
48. Kaya, I.: Two-degree-of-freedom IMC structure and controller design for integrating processes based on gain and phase-margin specifications. IEE Proc., Control Theory Appl. **154**(4), 481–487 (2004)
49. Kaya, I., Atherton, D.P.: A PI–PD controller design for integrating processes. In: Proceedings American Control Conference, pp. 258–262, San Diego, CA, 1999
50. Kaya, I., Tan, N., Atherton, D.P.: A simple procedure for improving performance of PID controllers. In: Proceedings IEEE International Conference on Control Applications, pp. 882–885, Istanbul, Turkey, 2003
51. Kempf, C.J., Kobayashi, S.: Disturbance observer and feedforward design for a high-speed direct-drive positioning table. IEEE Trans. Control Syst. Technol. **7**(5), 513–526 (1999)
52. Kharitonov, V.L., Niculescu, S.L., Moreno, J.: Static output feedback stabilization: necessary conditions for multiple delay controllers. IEEE Trans. Autom. Control **50**, 82–86 (2005)
53. Ko, B.-S., Edgar, T.F.: PID control performance assessment: the single-loop case. AIChE J. **50**, 1211–1218 (2004)
54. Kristiansson, B., Lennartson, B.: Robust and optimal tuning of PI and PID controllers. IEE Proc., Control Theory Appl. **149**(1), 17–25 (2001)
55. Kuehl, P., Horch, A.: Detection of sluggish control loops—experiences and improvements. Control Eng. Pract. **13**, 1019–1025 (2005)
56. Leva, A., Cox, C., Ruano, A.: Hands-on PID autotuning: a guide to better utilisation. Technical report, IFAC Technical Brief (2001), available at www.ifac-control.org
57. Lewis, F.L.: Optimal control. In: Levine, W.S. (ed.) The Control Handbook, pp. 759–778. CRC Press, Boca Raton (1996)
58. Li, H.X., Van Den Bosch, P.P.J.: A robust disturbance-based control and its application. Int. J. Control **58**(3), 537–554 (1993)
59. Liu, T., Gao, F.: Identification of integrating and unstable processes from relay feedback. Comput. Chem. Eng. **32**, 3038–3056 (2008)
60. Liu, T., Cai, Y.Z., Gu, D.Y., Zhang, W.D.: New modified Smith predictor scheme for integrating and unstable processes with time delay. IEE Proc., Control Theory Appl. **152**(2), 238–246 (2005)
61. Lu, X., Yang, Y.-S., Wang, Q.-G., Zheng, W.-X.: A double two-degree-of-freedom control scheme for improved control of unstable delay processes. J. Process Control **15**, 605–614 (2005)
62. Majhi, S., Atherton, D.P.: A new Smith predictor and controller for unstable and integrating processes with time delay. In: Proceedings IEEE International Conference on Decision and Control, pp. 1341–1345, Tampa, FL, 1998
63. Majhi, S., Atherton, D.P.: Modified Smith predictor and controller for processes with time delay. IEE Proc., Control Theory Appl. **146**(5), 359–366 (1999)
64. Majhi, S., Atherton, D.P.: Automatic tuning of the modified Smith predictor controllers. In: Proceedings IEEE International Conference on Decision and Control, pp. 1116–1120, Sydney, AUS, 2000
65. Majhi, S., Atherton, D.P.: Obtaining controller parameters for a new Smith predictor using autotuning. Automatica **36**, 1651–1658 (2000)
66. Majhi, S., Mahanta, C.: Tuning of controllers for integrating time delay processes. In: Proceedings IEEE International Conference on Electrical and Electronic Technology, pp. 317–320, Singapore, 2001

67. Marlin, T.E.: Process Control. McGraw-Hill, New York (1995)
68. Mascolo, S.: Congestion control in high-speed communication networks using the Smith principle. Automatica **35**(12), 1921–1935 (1999)
69. Matausek, M.R., Micic, A.D.: A modified Smith predictor for controlling a process with an integrator and long dead-time. IEEE Trans. Autom. Control **41**(8), 1199–1202 (1996)
70. Matausek, M.R., Micic, A.D.: On the modified Smith predictor for controlling a process with an integrator and long dead-time. IEEE Trans. Autom. Control **44**(8), 1603–1606 (1999)
71. Medvedev, A.: Disturbance attenuation in finite-spectrum-assignment. Automatica **33**(6), 1163–1168 (1997)
72. Mirkin, L., Zhong, Q.-C.: Coprime parametrization of 2DOF controller to obtain sub-ideal disturbance response for processes with dead time. In: Proceedings IEEE International Conference on Decision and Control, pp. 2253–2258, Orlando, USA, December 2001
73. Mirkin, L., Zhong, Q.-C.: 2DOF controller parametrization for systems with a single I/O delay. IEEE Trans. Autom. Control **48**(11), 1999–2004 (2003)
74. Mitchell, M.: An Introduction to Genetic Algorithms. MIT Press, Cambridge (1998)
75. Monroy-Loperena, R., Alvarez-Ramirez, J.: A note on the identification and control of batch distillation columns. Chem. Eng. Sci. **58**, 4729–4737 (2003)
76. Morari, M., Zafiriou, E.: Robust Process Control. Prentice-Hall, Inc., Englewood Cliffs (1989)
77. Nobuyama, E., Shin, S., Kitamori, T.: Deadbeat control of continuous-time systems: MIMO case. In: Proceedings IEEE International Conference on Decision and Control, pp. 2110–2113, Kobe, Japan, 1996
78. Normey-Rico, J.E., Camacho, E.F.: Robust tuning of dead-time compensators for process with an integrator and long dead-time. IEEE Trans. Autom. Control **44**(8), 1597–1603 (1999)
79. Normey-Rico, J.E., Camacho, E.F.: Smith predictor and modifications: a comparative study. In: Proceedings European Control Conference, Karlsruhe, Germany, 1999
80. Normey-Rico, J.E., Camacho, E.F.: A unified approach to design dead-time compensators for stable and integrative processes with dead-time. IEEE Trans. Autom. Control **47**(2), 299–305 (2002)
81. Normey-Rico, J.E., Camacho, E.F.: Control of Dead-time Processes. Springer, London (2007)
82. Normey-Rico, J.E., Camacho, E.F.: Unified approach for robust dead time compensator design. J. Process Control **19**(1), 38–47 (2009)
83. O'Dwyer, A.: Handbook of PI and PID Tuning Rules. Imperial College Press, London (2006)
84. Ogunnaike, B.A., Ray, W.H.: Process Dynamics, Modeling, and Control. Oxford University Press, New York (1994)
85. Ohishi, K., Nakao, M., Ohnishi, K., Miyachi, K.: Microprocessor-controlled DC motor for load-insensitive position servo system. IEEE Trans. Ind. Electron. **34**, 44–49 (1987)
86. Ou, L., Tang, Y., Gu, D., Zhang, W.: Stability analysis of PID controllers for integral processes with time delay. In: Proceedings American Control Conference, pp. 4247–4252, Portland, OR, 2005
87. Ou, L., Zhang, W., Gu, D.: Set of stabilising PID controllers for second-order integrating processes with time delay. IEE Proc., Control Theory Appl. **153**(5), 607–614 (2006)
88. Padula, F., Visioli, A.: On the stabilizing PID controllers for integral processes. Technical Report R.T. 2010-01-61, Department of Information Engineering, University of Brescia (2010)
89. Palmor, Z.J.: Time-delay compensation—Smith predictor and its modifications. In: Levine, S. (ed.) The Control Handbook, pp. 224–237. CRC Press, Boca Raton (1996)
90. Pao, L.Y.: Multi-input shaping design for vibration reduction. Automatica **35**(1), 81–89 (1999)
91. Patwardhan, R.S., Shah, S.L.: Issues in performance diagnostics of model-based controllers. J. Process Control **12**, 413–417 (2002)
92. Perez, H., Devasia, S.: Optimal output transitions for linear systems. Automatica **39**, 181–192 (2003)

93. Pfeiffer, B.-M.: Towards 'plug and control': self-tuning temperature controller for PLC. Int. J. Adapt. Control Signal Process. **14**, 519–532 (2000)
94. Piazzi, A., Visioli, A.: Minimum-time system-inversion-based motion planning for residual vibration reduction. IEEE/ASME Trans. Mechatron. **5**(1), 12–22 (2000)
95. Piazzi, A., Visioli, A.: Optimal inversion-based control for the set-point regulation of nonminimum-phase uncertain scalar systems. IEEE Trans. Autom. Control **46**, 1654–1659 (2001)
96. Piazzi, A., Visioli, A.: Optimal noncausal set-point regulation of scalar systems. Automatica **37**(1), 121–127 (2001)
97. Piazzi, A., Visioli, A.: Robust set-point constrained regulation via dynamic inversion. Int. J. Robust Nonlinear Control **11**, 1–22 (2001)
98. Piazzi, A., Visioli, A.: Using stable input-output inversion for minimum-time feedforward constrained regulation of scalar systems. Automatica **41**(2), 305–313 (2005)
99. Piazzi, A., Visioli, A.: A noncausal approach for PID control. J. Process Control **16**, 831–843 (2006)
100. Pontryagin, L.S.: On the zeros of some elementary transcendental functions. Am. Math. Soc. Transl. **2**, 95–110 (1955)
101. Poulin, E., Pomerleau, A.: PID tuning for integrating and unstable processes. IEE Proc., Control Theory Appl. **143**(5), 429–435 (1996)
102. Poulin, E., Pomerleau, A.: PI settings for integrating processes based on ultimate cycle information. IEEE Trans. Control Syst. Technol. **7**(4), 509–511 (1999)
103. Press, W.H., Teukolsky, S.A., Vetterling, W.T., Flannery, B.P.: Numerical Recipes: The Art of Scientific Computing. Cambridge University Press, Cambridge (1995)
104. Qin, S.J.: Control performance monitoring—a review and assessment. Comput. Chem. Eng. **23**, 173–186 (1998)
105. Quet, P.-F., Atalar, B., Iftar, A., Özbay, H., Kalyanaraman, S., Kang, T.: Rate-based flow controllers for communication networks in the presence of uncertain time-varying multiple time-delays. Automatica **38**(6), 917–928 (2002)
106. Seshagiri Rao, A., Rao, V.S.R., Chidambaram, M.: Set point weighted modified Smith predictor for integrating and double integrating processes with time delay. ISA Trans. **46**, 59–71 (2007)
107. Seshagiri Rao, A., Rao, V.S.R., Chidambaram, M.: Direct synthesis-based controller design for integrating processes with time delay. J. Franklin Inst. **346**, 38–56 (2009)
108. Rivera, D., Skogestad, S., Morari, M.: Internal model control. 4. PID controller design. Ind. Eng. Chem. Process Des. Dev. **25**(1), 252–265 (1986)
109. Schwartz, J.D., Rivera, D.E.: Control relevant demand modeling for supply chain management. In: Proceedings 14th IFAC Symposium of System Identification (SYSID), Newcastle, AUS, 2006
110. Shinskey, F.G.: Feedback Controllers for the Process Industries. McGraw-Hill, New York (1994)
111. Silva, G.J., Datta, A., Bhattacharyya, S.P.: New results on the synthesis of PID controllers. IEEE Trans. Autom. Control **47**, 241–252 (2002)
112. Silva, G.J., Datta, A., Bhattacharyya, S.P.: PID Controllers for Time Delay Systems. Birkhäuser, Boston (2005)
113. Skogestad, S.: Simple analytic rules for model reduction and PID controller tuning. J. Process Control **13**, 291–309 (2003)
114. Smith, O.J.M.: Feedback Control Systems. McGraw-Hill, New York (1958)
115. Srividya, R., Chidambaram, M.: On line controllers tuning for integrator plus delay systems. Process Control Qual. **9**, 59–66 (1997)
116. Sung, S.W., Lee, I.: Limitations and countermeasures of PID controllers. Ind. Eng. Chem. Res. **35**, 2596–2610 (1996)
117. Sung, S.W., Lee, I.-B., Lee, B.-K.: On-line process identification and automatic tuning method for PID controllers. Chem. Eng. Sci. **53**, 1847–1859 (1998)
118. Swanda, A.P., Seborg, D.E.: Controller performance assessment based on set-point response data. In: Proceedings American Control Conference, pp. 3863–3867. San Diego, CA, 1999

119. Thyagarajan, T., Yu, C.-C.: Improved autotuning using the shape factor from relay feedback. Ind. Eng. Chem. Res. **42**, 4425–4440 (2003)
120. Tian, Y.-C., Gao, F.: Control of integrator processes with dominant time delay. Ind. Eng. Chem. Res. **38**, 2979–2983 (1999)
121. Tsypkin, Y.Z.: Robust internal model control. ASME J. Dyn. Syst., Meas. Control **115**(2B), 419–425 (1993)
122. Tyreus, B.D., Luyben, W.L.: Tuning PI controllers for integrator/dead time processes. Ind. Eng. Chem. Res. **31**, 2625–2628 (1992)
123. Umeno, T., Hori, Y.: Robust speed control of dc servomotors using modern two degrees-of-freedom controller design. IEEE Trans. Ind. Electron. **38**(5), 363–368 (1991)
124. Varaiya, P., Walrand, J.: High-performance Communication Networks. Morgan Kaufmann, San Francisco (1996)
125. Veronesi, M., Visioli, A.: A technique for abrupt load disturbance detection in process control systems. In: Proceedings 17th IFAC World Congress on Automatic Control, pp. 14900–14905, Seoul, ROK, 2008
126. Veronesi, M., Visioli, A.: Performance assessment and retuning of PID controllers. Ind. Eng. Chem. Res. **48**, 2616–2623 (2009)
127. Veronesi, M., Visioli, A.: Performance assessment and retuning of PID controllers for integral processes. J. Process Control **20**, 261–269 (2010)
128. Visioli, A.: Optimal tuning of PID controllers for integral and unstable processes. IEE Proc., Control Theory Appl. **148**(2), 180–184 (2001)
129. Visioli, A.: Time-optimal plug&control for integrating and FOPDT processes. J. Process Control **13**, 195–202 (2003)
130. Visioli, A.: A new design for a PID plus feedforward controller. J. Process Control **14**, 455–461 (2004)
131. Visioli, A.: Method for proportional-integral controller tuning assessment. Ind. Eng. Chem. Res. **45**, 2741–2747 (2006)
132. Visioli, A.: Practical PID Control. Springer, London (2006)
133. Visioli, A.: Experimental evaluation of a time-optimal plug&control strategy. ISA Trans. **46**, 519–525 (2007)
134. Visioli, A., Piazzi, A.: Improving set-point following performance of industrial controllers with a fast dynamic inversion algorithm. Ind. Eng. Chem. Res. **42**, 1357–1362 (2003)
135. Wallen, A.: Tools for autonomous process control. PhD thesis, Lund Institute of Technology, Lund (2000)
136. Wallen, A., Åström, K.J.: Pulse-step control. In: Proceedings 15th IFAC World Congress on Automatic Control, Barcelona, E, 2002
137. Wang, B., Rees, D., Zhong, Q.-C.: Control of integral processes with dead time. Part IV: various issues about PI controllers. IEE Proc., Control Theory Appl. **153**(3), 302–306 (2006)
138. Wang, L., Barnes, T.J.D., Cluett, W.R.: New frequency-domain design method for PID controllers. IEE Proc., Control Theory Appl. **142**, 265–271 (1995)
139. Wang, L., Cluett, W.R.: Tuning PID controllers for integrating processes. IEE Proc., Control Theory Appl. **144**(5), 385–392 (1997)
140. Wang, Q.-G., Fung, H.-W., Zhang, Y.: Robust estimation of process frequency response from relay feedback. ISA Trans. **38**, 3–9 (1999)
141. Wang, Y.-G., Cai, W.-J.: PID tuning for integrating processes with sensitivity specification. In: Proceedings IEEE International Conference on Decision and Control, pp. 4087–4091, Orlando, FL, 2001
142. Wang, Y.-G., Cai, W.-J.: Advanced proportional-integral-derivative tuning for integrating and unstable processes with gain and phase margin specifications. Ind. Eng. Chem. Res. **41**, 2910–2914 (2002)
143. Wang, Y.-G., Cai, W.-J., Shi, Z.-G.: PID autotuning for integrating processes with specifications on gain and phase margins. In: Proceedings American Control Conference, pp. 2181–2185, Arlington, VA, 2001
144. Warburton, R.D.H., Disney, S.M., Towill, D.R., Hodgson, J.P.E.: Further insights into "the stability of supply chains". Int. J. Prod. Res. **42**(3), 639–648 (2004)

145. Watanabe, K., Ito, M.: A process-model control for linear systems with delay. IEEE Trans. Autom. Control **26**(6), 1261–1269 (1981)
146. Watanabe, K., Nobuyama, E., Kojima, A.: Recent advances in control of time delay systems—a tutorial review. In: Proceedings IEEE International Conference on Decision and Control, pp. 2083–2089, Kobe, Japan, 1996
147. Weiss, G., Hafele, M.: Repetitive control of MIMO systems using H^∞ design. Automatica **35**(7), 1185–1199 (1999)
148. Wu, C.: Intelligent use of delayed information in the supply chain by artificial neural network. In: Proceedings IEEE International Conference on Systems, Man, and Cybernetics, vol. 2, pp. 66–70 (1999)
149. Xu, J., Shao, H.: Advanced PID tuning for integrating processes with a new robustness specification. In: Proceedings American Control Conference, pp. 3961–3966, Denver, Colorado, 2003
150. Xu, J., Shao, H.: A novel method of PID tuning for integrating processes. In: Proceedings IEEE International Conference on Decision and Control, pp. 139–142, Maui, Hawaii, 2003
151. Yu, C.-C.: Autotuning of PID Controllers: Relay Feedback Approach. Springer, London (1999)
152. Zhang, M., Jiang, C.: Problem and its solution for actuator saturation of integrating process with dead time. ISA Trans. **47**, 80–84 (2008)
153. Zhang, W., Sun, Y.X.: Modified Smith predictor for controlling integrator/time delay process. Ind. Eng. Chem. Res. **35**, 2769–2772 (1996)
154. Zhang, W., Xu, X., Sun, Y.: Quantitative performance design for integrating processes with time delay. Automatica **35**, 719–723 (1999)
155. Zhang, W., Rieber, J.M., Gu, D.: Optimal dead-time compensator design for stable and integrating processes with time delay. J. Process Control **18**, 449–457 (2008)
156. Zhong, Q.-C.: Control of integral processes with dead-time. Part 3: Deadbeat disturbance response. IEEE Trans. Autom. Control **48**(1), 153–159 (2003)
157. Zhong, Q.-C.: Robust stability analysis of simple systems controlled over communication networks. Automatica **39**(7), 1309–1312 (2003)
158. Zhong, Q.-C.: On distributed delay in linear control laws. Part I: Discrete-delay implementations. IEEE Trans. Autom. Control **49**(11), 2074–2080 (2004)
159. Zhong, Q.-C.: On distributed delay in linear control laws. Part II: Rational implementations inspired from the δ-operator. IEEE Trans. Autom. Control **50**(5), 729–734 (2005)
160. Zhong, Q.-C.: Robust Control of Time-delay Systems. Springer, Berlin (2006)
161. Zhong, Q.-C., Li, H.X.: Two-degree-of-freedom PID-type controller incorporating the Smith principle for processes with dead-time. Ind. Eng. Chem. Res. **41**(10), 2448–2454 (2002)
162. Zhong, Q.-C., Mirkin, L.: Control of integral processes with dead-time. Part 2: Quantitative analysis. IEE Proc., Control Theory Appl. **149**(4), 291–296 (2002)
163. Zhong, Q.-C., Normey-Rico, J.E.: Control of integral processes with dead-time. Part 1: Disturbance observer-based 2DOF control scheme. IEE Proc., Control Theory Appl. **149**(4), 285–290 (2002)
164. Zhong, Q.-C., Xie, J.Y., Jia, Q.: Time delay filter-based deadbeat control of process with dead time. Ind. Eng. Chem. Res. **39**(6), 2024–2028 (2000)
165. Zhou, K., Doyle, J.C.: Essentials of Robust Control. Prentice-Hall, Upper Saddle River (1998)
166. Ziegler, J.G., Nichols, N.B.: Optimum setting for automatic controllers. ASME Trans. 759–768 (1942)
167. Zou, Q., Devasia, S.: Preview-based stable-inversion for output tracking of linear systems. ASME J. Dyn. Syst. Meas. Control **121**, 625–630 (1999)

Index

A. Visioli, Q.-C. Zhong, *Control of Integral Processes with Dead Time*,
Advances in Industrial Control,
DOI 10.1007/978-0-85729-070-0,

© Springer-Verlag London Limited 2011

Other titles published in this series (continued):

Soft Sensors for Monitoring and Control of Industrial Processes
Luigi Fortuna, Salvatore Graziani, Alessandro Rizzo and Maria G. Xibilia

Adaptive Voltage Control in Power Systems
Giuseppe Fusco and Mario Russo

Advanced Control of Industrial Processes
Piotr Tatjewski

Process Control Performance Assessment
Andrzej W. Ordys, Damien Uduehi and Michael A. Johnson (Eds.)

Modelling and Analysis of Hybrid Supervisory Systems
Emilia Villani, Paulo E. Miyagi and Robert Valette

Process Control
Jie Bao and Peter L. Lee

Distributed Embedded Control Systems
Matjaž Colnarič, Domen Verber and Wolfgang A. Halang

Precision Motion Control (2nd Ed.)
Tan Kok Kiong, Lee Tong Heng and Huang Sunan

Optimal Control of Wind Energy Systems
Julian Munteanu, Antoneta luliana Bratcu, Nicolaos-Antonio Cutululis and Emil Ceangă

Identification of Continuous-time Models from Sampled Data
Hugues Gamier and Liuping Wang (Eds.)

Model-based Process Supervision
Arun K. Samantaray and Belkacem Bouamama

Diagnosis of Process Nonlinearities and Valve Stiction
M.A.A. Shoukat Choudhury, Sirish L. Shah and Nina F. Thornhill

Magnetic Control of Tokamak Plasmas
Marco Ariola and Alfredo Pironti

Real-time Iterative Learning Control
Jian-Xin Xu, Sanjib K. Panda and Tong H. Lee

Deadlock Resolution in Automated Manufacturing Systems
ZhiWu Li and MengChu Zhou

Model Predictive Control Design and Implementation Using MATLAB®
Liuping Wang

Predictive Functional Control
Jacques Richalet and Donal O'Donovan

Fault-tolerant Flight Control and Guidance Systems
Guillaume Ducard

Fault-tolerant Control Systems
Hassan Noura, Didier Theilliol, Jean-Christophe Ponsart and Abbas Chamseddine

Detection and Diagnosis of Stiction in Control Loops
Mohieddine Jelali and Biao Huans (Eds.)

Stochastic Distribution Control System Design
Lei Guo and Hong Wang

Dry Clutch Control for Automotive Applications
Pietro J. Dolcini, Carlos Canudas-de-Wit and Hubert Béchart

Advanced Control and Supervision of Mineral Processing Plants
Daniel Sbarbaro and Ren6 del Villar (Eds.)

Active Braking Control Design for Road Vehicles
Sergio M. Savaresi and Mara Tanelli

Active Control of Flexible Structures
Alberto Cavallo, Giuseppe de Maria, Ciro Natale and Salvatore Pirozzi

Induction Motor Control Design
Riccardo Marino, Patrizio Tomei and Cristiano M. Verrelli

Fractional-order Systems and Controls
Concepcion A. Monje, YangQuan Chen, Blas M. Vinagre, Dingyu Xue and Vincente Feliu

Model Predictive Control of Wastewater Systems
Carlos Ocampo-Martinez

Tandem Cold Metal Rolling Mill Control
John Pittner and Marwan A. Simaan
Publication due December 2010

Drives and Control for Industrial Automation
Kok Kiong Tan and Andi Sudjana Putra
Publication due December 2010

MIX
Papier aus verantwortungsvollen Quellen
Paper from responsible sources
FSC® C105338

If you have any concerns about our products,
you can contact us on
ProductSafety@springernature.com

In case Publisher is established outside the EU,
the EU authorized representative is:
Springer Nature Customer Service Center GmbH
Europaplatz 3, 69115 Heidelberg, Germany

Printed by Libri Plureos GmbH
in Hamburg, Germany